T0173874

PROCEEDINGS OF THE 30TH INTERNATIONAL GEOLOGICAL CONGRESS
VOLUME 4

STRUCTURE OF THE LITHOSPHERE AND DEEP PROCESSES

Proceedings of the 30th International Geological Congress

Proceedings of the
30th International Geological Congress

Beijing, China, 4 - 14 August 1996

VOLUME 4

Structure of the Lithosphere and Deep Processes

Editor:
Hong Dawei
Research Centre of Lithosphere, Institute of Geology, Beijing, China

CRC Press
Taylor & Francis Group
Boca Raton London New York

CRC Press is an imprint of the
Taylor & Francis Group, an **informa** business

First published 1997 by VSP BV

Published 2019 by CRC Press
Taylor & Francis Group
6000 Broken Sound Parkway NW, Suite 300
Boca Raton, FL 33487-2742

© 1997 by Taylor & Francis Group, LLC
CRC Press is an imprint of Taylor & Francis Group, an Informa business

First issued in paperback 2019

No claim to original U.S. Government works

ISBN 13: 978-0-367-44810-3 (pbk)
ISBN 13: 978-90-6764-261-3 (hbk)

Visit the Taylor & Francis Web site at
http://www.taylorandfrancis.com

and the CRC Press Web site at
http://www.crcpress.com

CONTENTS

PREFACE

16 papers in this volume are representative of themes covered during symposium on"Structure of the lithosphere and deep processes" held at the 30th International Geological Congress in Beijing, China during August, 1996.

The symposium was divided into 7 sessions which covered following aspects: the nature of the lower crust and exchange between the crust and the mantle; three-dimensional models and four-dimensional mapping of the lithosphere; anisotropy of the mantle and mantle rheology; terrestrial heat flow, hot spots, mantle plumes, and thermal structure of the lithosphere; the global geoscineces transects; continental scientific drilling and site selection for scientific drilling in China; lithospheric structure, composition, and dynamics of orogenic belts.

The 16 papers in this volume have been selected to represent different aspects of research discussed during the 30th IGC. For reasons of length and cost, authors have been limited to presenting only the essential results necessary to develop their arguments and ideas.

These papers provide a flavor of current reseach into the structure of the lithophere and deep processes. I hope these papers provide a stimulus to further exciting reseach in this important area of earth sciences.

I would particularly like to thank the authors, who all managed to supply their final typescripts so promptly after the congress, and Jinfu Deng, Jiwen Teng, Xianjie Shen, Guodong Liu, H.-J. Gotze, Guangzhi Liu and Rui Feng who assisted in various way in the preparation of this book.

Dawei Hong

Proc. 30 *Int'l. Geol. Congr.*, Vol. 4, pp. 1-10
Hong Dawei (Ed)

Nature of Lower Crust and Crust-mantle Exchange

DENG JINFU, WU ZONGXU, ZHAO HAILING, LUO ZHAOHUA, CAO YONYQING
China University of Geosciences, Beijing 100083, China

Abstract

Based on the petrological approach in combination with geophysical model and experimental or calculated data of Vp & Vs for the rocks, the crust-mantle petrological structure of three geotectonic units of china are presented by this paper . From oceanic crust through continental collision belt, with rapid increasing crust thickness, the mineral facies of the lower crust is changed from normal igneous or green-schist facies through amphibolite and granulite facies to eclogite facies, in correspondence with which the composition is changed from basaltic through andesitic to granitic, the compositional trend is called "granitization" resulted from the basaltic eclogite crust recycling back into the mantle .However, from the continental collision belt to the craton or through continental rifting to the oceanic crust, the crust thickness, mineral facies and composition of the lower crust are developed oppsitely, the compositional trend is called "basification" generated by the basaltic magma underplating into the lower crust from the mantle .

Keywords: lower crust, crust-mantle exchange, petrological structure, granitization, basification

INTRODUCTION

A consensus view is that new oceanic crust is formed at mid-ocean ridge, the crust is composed of three layer, i.e. layer1—thin sedimental cover, layer2—basaltic lava, layer3—gabbro and diabase. Continental crust is believed to generate mainly in the Precambrian by the formation of continental nuclei and craton .Juvenile continental crust in the Phanerozoic is regarded to be formed at island-arc and active continental margin. However, some questions related to the nature and formation of the continental crust, especially the lower crust, remain to be open .(1) The two layer model of continental crust structure has been substituted by the three-layer model[15]. The three layer model of continental crust has been constructed from the regions of normal and thinned crust with 30-40km thickness. Is the three layer continental crust model suitable for the thickened crust area with 70-80km thickness like the Qinghai-Tibet plateau? (2)The average composition of total crust is similar to andesite or dacite[3,21,23].However, the Juvenile continental crust materials generated from the mantle are basaltic composition. How to explain this difference? (3) It is generally considered the upper and lower continental crust to be granitic and gabbroic(or dioritic) composition, respectively. However, the exposed upper and lower (average granulite-facies terrane)crust on the Canadian Shield[3]are similar each to other as granitic or dacite composition. Is the lower continental crust different or similar to the upper crust in the composition ?(4)Is the crustal composition for the different tectonic setting different or similar?(5)Does the crust petrology change during the crustal thickening or thinning processes? (6)What are the processes to control the nature of the lower crust ?(7)Where is the locality to take the mass exchange between the crust and the mantle? How to realize the mantle-crust exchange ?We try to discuss these questions, nature of lower crust, and crust-mantle exchange, taking the China continent for example, and then compare the continental crust to the oceanic crust and island-arc crust.

APPROACH TO CRUST-MANTLE PETROLOGICAL STRUCTURE

In a similar way to the P-wave velocity(Vp) structure or the density structure, the petrological structure[27]has been suggested in order to emphasis on study of crust and mantle using petrological method. Petrological approach for study on the crust and the mantle structure comes mainly from three aspect as follows: (1)exposed deep crust rocks or even cross-section of the crust, mainly the Precambrian metamorphic rocks; (2)crust and mantle xenoliths and fragments brought to the surface by the magmatic or tectonic processes; (3)chemical and physical information of magma source. About 70-90% of the present crustal mass has been formed by about 2500Ma ago[21].After the formation of Archean metamorphic rocks, the continental crustal mass and its structure often reworked and modified by the later magmatic and tectonic events, which must be considered for the construction of the petrological structure .Comprehensive study from above mentioned aspects can be resulted in the construction of crust-mantle petrological structure, in which the petrological phase equilibrium and geothermobarometry as well as the mineral-silicate melt equilibrium thermodynamics investigation are thought to be the key to modelling of the petrological structure [9].

Geophysical approach includes mainly the Vp and Vs data. Experimental measurements of Vp and Vs for mineral and rock are linkage between the petrological and geophysical approach .Petrological approach in combination with geophysical model gives the best constraints for crust-mantle petrological structure [15,8,24]

CRUST-MANTLE PETROLOGICAL STRUCTURE

As consistent with the geotectonic units on a lithospheric scale [14]three type of the crust-mantle petrological structure of China continent may be recognized :continental rifting type of East China ;cratonic type of Middle China and intracontinental orogenic type of Qinghai-Tibet-Himalaya, as shown in Table 1&2.

Fountain's three layer structure model of continental crust[15], based mainly on Vp-data and petrological studies, has gained widespread acceptance[16,24], i.e. the upper, middle and lower continental crust are composed of greenschist, amphibolite and granulite facies rocks, respectively. The continental crust with three-layer structure may be divided into continental rifting and cratonic types, as shown in table1[27]. Recent research shows that the continental crust of Qinghai-Tibet-Himalaya intracontinental orogenic belt has four layer structure (Tab.2)[12,13]. In order to be compare with three layer model, we call the forth layer a thickened lower crust including mountain root.

Based on both p-t condition of the phase transition boundaries of granulite and eclogite facies [4,17,28], and the geotherm of several localities of Qinghai-Tibet-Himalaya orogenic belt [19,20], we may suggest that at the depth of about 35-40km, Opx-out, high-pressure

Table 1. Crust-mantle petrological structure model(simplified after Wu et al.[27])

continental rifting zone (North China)

H (km)	Layer		Petro-logy	V_p	V_s (km/s)
0	Crust	U.	Chl.F. γδ	6.1	3.45
20		M.	Am.F. γδ	6.4 / 6.1	3.75 / 3.62
		L.	Py.F. δ b	6.5 / 7.2	3.80
40	Upper mantle		Sp-Lherzolite (Harzburgite Dunite Pyroxenite)	7.9 ↓ 8.1	4.45
60			L/A		
80			Lherzolite with ≈7% basaltic melt	7.6	4.20
100			Ga-Lherzolite & Harzburgite	8.25	4.12

Craton (Erdos)

H (km)	Layer		Petro-logy	V_p	V_s (km/s)
0	Crust	U.	Chl.F. γ	5.8	3.55
20		M.	Am.F. γ	6.1	3.68
		L.	Py.F. γδ	6.4	3.8
40			H.P.Py.F. γδ	6.7	4.40
60	Upper mantle		Sp-Harzburgite & Dunite	8.0 ↓ 8.25	4.40
80					
100			Ga-Harzburgite & Dunite L/A≈200km	8.35	4.20

Notes: U-upper, M-middle, L-lower, L/A-boundary between lithosphere and asthenosphere, chl.F-greenschist facies, Am.F.-amphibolite facies, Py.F.-granulite facies, H.P.Py.F:- high pressure granulite facies

Table 2. Crust-mantle petrological structure model, Qinghai-Tibet-Himalaya intracontinental orogenic belt (simplified after Deng et al.[14])

without magma underplating (Qilian & Himalaya)

H (km)	Layer	Petro-logy	V_p (km/s)	V_s (km/s)
Crust	U.	Chl.F. γ	5.7 / 6.1	3.45
	M.	Am.F. γ	5.8	3.65
	L.	Py.F. γ	6.4	3.75
	T.L.	H.P.Py.F. $\gamma\delta$		3.85
			6.8	
	M.R.	Eclog.F. γ		4.15
		Moho		
Upper mantle		Basaltic eclogite cover		
		Ga-Peridotite L/A≈150km (Qilian)	8.1	4.45

with magma underplating (Bayan Har & Gangdise)

H (km)	Layer	Petro-logy	V_p (km/s)	V_s (km/s)
Crust	U.	Chl.F. γ	4.7 / 6.0	2.64 / 3.43
	M.	Am.F. γ	5.75	3.85
	L.	Py.F. γ	6.3	3.33
	T.L.	H.P.Py.F. γ	6.5	3.80
	M.R.	Eclog.F. δ	7.4	3.98 / 4.56
		Moho		
Upper mantle		Basaltic eclogite cover		
		Ga-Peridotite L/A≈200km (Bayan Har) ≈120km (Gangdise)	8.15	4.65

Note : T.L.-thickened lower crust, M.R.-mountain root, Eclog.F.-eclogite facies, Others see notes of Tab.1

granulite facies (HPGF) is formed, at the depth of about 45-60km, Pl-out, eclogite facies is formed. So we suggest a four-layer structure model for the thickened continental crust rather than the Fountain's three-layer structure model . The forth layer bellow about 40km is called a thickened lower crust, in general, and includes mountain root when the depth is greater than 45-60km and the eclogite facies is reached[13,14].

It is well known that the continental crust is composed mainly of quartz and feldspar. Experimental measurements [1,2] show that quartz has lower Vp, but higher Vs, however, feldspar has lower Vp, but higher Vs. Therefore, we introduce both the Vp and Vs in order to construct the best model.From the Vp and Vs of minerals, the Vp and Vs of rocks may be calculated[13,14], which is consistent, in general, with the experimental measurements[1,2,15,18,25].

It is suggested that the thickened lower crust including mountain root are composed of granitic composition by a comparison of the Vp and Vs data between the geophysical model (Tab.2) and the caculated as well as experimental measurements.

It is of interest to note from Tab.1 that a lower Vs is occurred at about the depth of 80km for the cratonic block(Erdos), but there is no lower Vp at the same depth. The lower Vs in the upper mantle is regarded as a indication of the asthenosphere by some geologists, but the Vp does not indicate the asthenosphere.A simple calculation based on the Vp & Vs data of minerals shows that during transition from spinel to garnet peridotite ΔVp =+0.199km/s, ΔVs=-0.227km/s, from which the lower Vs with the higher Vp at the depth of about 80km is considered to be resulted from the phase transition[27].

NATURE OF LOWER CRUST.

Table 1 shows a big difference of the lower crust in petrology, Vp and Vs between cratonic block and continental rifting zone. For the continental rifting zone the petrological composition of the lower crust is dioritic, at the base of the crust is basaltic. For the cratonic block the lower crust has granodioritic composition. Cenozoic basaltic volcanism containing upper mantle peridotite xenoliths is widespread distributed in the continental rifting zone of East China[6,5,11]. Petrological study and mineral-silicate melt equilibrium thermodynamic calculations show that the megacryst assemblage of clinopyroxene, anorthoclase and garnet is widespread developed in the Cenozoic basalts, and the clinopyroxene -anorthoclase -garnet cumulate is formed from the magmatic chamber situated at the base of the crust[7,8]. The basic granulite of basaltic composition and clinopyroxene -anorthoclase -garnet cumulate are consistent with the high Vp at the base of the crust. So we may assume that the basic composition at the base of the crust for the East continental rifting zone is resulted from the basaltic magma underplating. As compared to the granitic composition of the lower crust for the cratonic block, the composition of the lower crust for the East China shows a "basification" of the continental crust during continental rifting due to the basaltic magma underplating.

Qinghai-Tibet-Himalaya intracontinental orogenic belt has two subtypes of the crust-mantle petrological structure : with magmatic underplating for Bayan Har and Gangdise, and without magmatic underplating for Qilian and Himalaya(Tab.2), the former has mountain root with higher Vp and corresponds to volcanic eruptions of shoshonitic or andensitic series at the surface since Cenozoic, the latter has a mountain root of granitic composition and corresponds to no volcanic eruptions at the surface in Qilian or only the muscovite /two mica granites related to the intracontinental subduction and no related to the magmatic underplating from the mantle in Himalaya since Cenozoic [14]. It is obviously from Tab.2 that the average composition of the continental crust is granitic and the mountain root has no basaltic composition regardless the crust-mantle petrological structure with or without magmatic underplatng. However, the geologic studies show that before the collisional orogeny the crust must be contained basaltic rocks, the Cenozoic volcanic eruptions of shoshonitic or andensitic series suggest that the basaltic underplating from the mantle must be occurred in Cenozoic . So, a question is arisen, where do the basaltic materials go away from the crust? What is the mechanism?

We suggested a model shown in Fig.1 in order to answer the above-mentioned question[13].

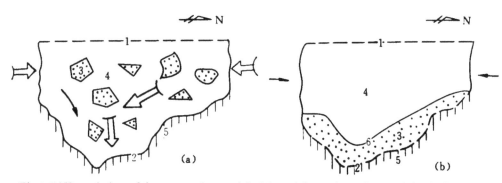

Fig.1 Differentiation of deep crustal materials (a), and formation of new Moho(b),[13]

1. Lower boundary of the middle crust, 2. Paleo-Moho, 3. Eclogite of basaltic composition, 4. Eclogite and high-pressure granulite of granitic composition, 5. Peridotite, 6. New Moho.

From the p-t condition, the eclogite in the basaltic composition is formed at shallower level than the eclogite in granitic composition. In the continental crust composed mainly of granitic materials, 500 °C-isotherm corresponds generally to the boundary between plastic and brittle deformation, and the boundary is located at about the depth of 20-25km [26]. At the crustal environment, when the granitic lower crust behaves a plastic nature, the basaltic lower crust behaves still more brittle nature. During the processes of crust thickening, the granitic crust beneath the base of the middle crust is formed a plastic flow, but the basaltic rock may be ruptured and descend downwards together with the granitic plastic flow(Fig.1a), at the same time the granitic and basaltic material are transited through high pressure granulite to eclogite at different depth. When the compressional stress weakens, the basaltic eclogite fragment must descend downwards separated from the granitic plastic flow and accumulated above the paleo-Moho due to the higher density than that of the granitic high-pressure granulite and eclogite(Fig.1b). Because the density and seismic velocity difference between the basaltic eclogite and upper mantle peridotite is much more lower than that between the basaltic eclogite and the granitic eclogite, the interface for the sudden change of the seismic velocity and density should be moved to the boundary between the basaltic eclogite cumulate and the granitic eclogite and high-pressure granulite, hence the new Moho is formed (Fig.1b) between the granitic or dioritic eclogite in the mountain root and the basaltic eclogite cover in the uppermost of the upper mantle. So, we may say, during the orogenic processes a "granitization" of the continental crust may be resulted from the basaltic eclogite crust recycling back into the mantle.

Using the cratonic crust-mantle petrological structure as a reference frame(Tab.1), it is obviously that the thickness of the upper and middle crust is approximately the same for the three tectonic units in China(Tab.1,2), but the thickness of the lower crust, including the thickened lower crust and mountain root, is much more different between the thinned and thickened lower crust for the East rifting zone with extensional stress and the Qinghai-Tibet-Himalaya orogenic belt with compressional stress, respectively, from which the plastic nature of the continental lower crust, including the thickened lower crust and mountain root, is suggested generally. The earthquake hypocenter of the continent is mainly concentrated in the depth of 10-20km,which suggests a brittle and rigid nature of the shallow crust, and a ductile and plastic nature of the deep crust. Experimental data show that the boundary between brittle and ductile behavior for the granitic rock is about 500 ℃ and earthquake hypocenter is situated above the 500 ℃isotherm in the continental crust, which suggests also the brittle and ductile nature for the shallow and deep crust, respectively.

Form above-mentioned we can reach a conclusion, the tectonic environment is a fundamental factor controlling the mineralogical and petrological composition, nature and thickness of the shallow and deep continental crust.

CRUST-MANTLE EXCHANGE AND CRUSTAL EVOLUTION

The area adjacent to the crust-mantle interface (the Moho) is considered to be a best and major locality of mantle-crust exchange, where the basaltic magma underplating from mantle into crust and the basaltic eclogite recycling back into the mantle from crust are taking place. Oceanic subduction and intracontinental subduction are regarded as the major mechanism for the shallow crust recycling back into mantle and deep crust, respectively. The deep crustal plastic mass flow within the continent is thought to be a best environment for the gravitational differentiation of crustal material in solid state. The underplating magma pond(or sea) at the base of continental crust is considered to be a best locality for the mass and heat exchange between crust and mantle, as well as for the crust-mantle magma mixing, crystal and gravitational differentiation. The high Vp and Vs layer at the base of the crust, resulted from the magmatic underplating, is about 7-8km and 20km in thickness for East China, and Bayan Har & Gangdise, respectively(Tab.1,2), from which we suggest that about 20-25% of the new crustal material generated from the mantle is added and contributed to the total crust thickness. As viewed from the geological time scale, during a short time since Cenozoic such a large fluxes of mantle materials adding into the crust is enough to show a great effect of the lithospheric thinning as dynamic source for tectonic activity.

Petrological phase equilibrium and experiments show that the basaltic and granitic magma are represented the minimum-melt composition of the upper mantle and the crustal materials system, respectively. It is well know that the basaltic material is regarded as a new or juvenile crust resulted from the upper mantle, and the granitic material are thought as a secondary crust generated from the basaltic and dioritic crust materials. We may regard the granitic crust to be a end product of differentiation of the crust-mantle material system, and the basaltic crust to be an initial product of differentiation of crust-mantle material system.

From oceanic crust through island-arc and continental margin-arc to continental collisional orogenic belt, the crustal thickness is thickened rapidly, the mineral facies of the lower crust is changed from normal igneous or greenschist facies through the amphibolite-granulite facies to eclogite facies, in correspondence with which the composition evolved from basaltic through andesitic to granitic, this evolutionary trend represents the processes of differentiation of crust-mantle materials. However, during crustal thinning, beginning of the collapse of intracontinental orogenic belt, two evolutionary directions are developed, one is formed the crust with normal thickness and turned into a tectonically stable craton, the other is continued to be crustal thinning, and the continental rifting, even the continental marginal sea and oceanic crust are resulted in. With the crustal thinning, the mineral facies and petrological composition are developed oppositely, from granitic through andesitic (granitic +basaltic)to oceanic basaltic crust, this is a process of crustal "basification", representing the results of the new materials adding to the crust generated from the mantle differentiation.

In general, the process of crust-mantle material differentiation from basaltic to granitic might be compared to that one basaltic composition divides into two end member component, the felsic end member (granitic) remained behind in the crust, the ultramafic end member recycling back into the mantle, therefore, "granitization" of the continental crust is regarded to be a recycling process of the crustal ultramafic and mafic material back into the mantle. In contrast, the "basification" of the crust is a process of the mantle materials adding to the crust, this is a process of mass mixing for the continental crust, which is a opposite process from the mass differentiation process. Fig.2 shows the above-mentioned process and the crust evolution.

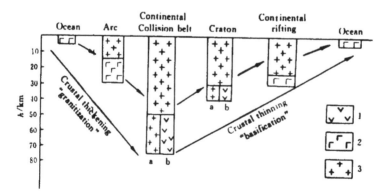

Fig. 2. Cartoon showing the crustal evolution

1. Basaltic, 2. Andesitic, 3. Granitic

Since the Earth has been shaped the plate-tectonic processes, the basaltic magma derived from the mantle by the partial melting processes is ascended or along the mid-oceanic ridge forming oceanic crust, or into the continent and underplated at the base of the crust. Oceanic crust, then, recycling back into the mantle by the subduction processes. From the underplating basaltic magma chamber at the base of the continental crust, the mafic component may be recycled back into the mantle by the crystal fractionation, and granitic and diortic component remains behind in the crust, or by the solid basaltic materials recycling back into the mantle, the granitization of the continental crust takes place during orogenic processes. The way of Archaen crustal formation may be different from the Phanerozoic, but the trend of the mass differentiation may be the same as judged by the fact that the Archaen tonalite-trondhjemite crust (corresponding to andesite, in general) has been evolved into the Proterozoic granodriorite-granite crust (corresponding to granite).

It seems that the continental grantic end member is formed more and more, and the crustal composition is departed more and more from the mantle with progressive differentiation of the crust-mantle materials. With the mixing processes of the crust-mantle materials, the oceanic crust is formed finally, and the crustal composition is closed to the mantle. Spiral cycling of the crust-mantle exchange takes place through both the differentiation and mixing processes of the crust-mantle materials, however, increasing of the grantic continent might be a general tendency with time.

Acknowledgments

This work is supported by both the National Natural Science Foundation of China and the Science Foundation of the Ministry of Geology and Mineral Resources of China.

REFERENCES

* in Chinese with English summary, others in English

1.   F.Birch. The velocity of compressional waves in rocks to 10kb, Jour. *Geophys. Res.* **66**, 2199-2224(1961).
2.   N.I.Christernsen and D.M. Fountain. Constitution of the lower continental crust based on experimental studies of seismic velocity in granulite, *Geol. Soc. Amer. Bull.* **86**, 227-236 (1975).
3.   K.C.Condie. *Plate tectonics and crustal evolution*, Pergrmon Press (1982).
4.*   Jinfu Deng. *Petrological phase diagram and petrogenesis*, Wuhan College of Geology Press (1987).
5.*   Jinfu Deng. Continental rifting magmatism and deep processes. In: *Study on Cenozoic basalts and upper mantle in East China.* J.C. Chi(Ed.). pp. 201-218. China University of Geosciences Press (1988).
6.   Jinfu Deng, Molan E and Fengxian Lu. Neozoic basalts in northeast China and their relation to continental rift tectonics, *27th IGC. Abst.* IV(08,09), 289-290 (1984).
7.*   Jinfu Deng, Molan E and Fengxian Lu.The composition, structure and thermal condition of the upper mantle beneath northeast China, *Acta Petrol. Miner.* **6**,1-10 (1987).
8.*   Jinfu Deng, Fengxian Lu and Molan E.The origin and ascending p-t path of the Hannuoba basaltic magma, *Geol. Rev.* **33**, 317-324 (1987).

9.*   Jinfu Deng, Delong Ye, Hailing Zhao and Deping Tang. *Volcanism, deep processes and basin formation in the lower reaches of the Yangtze River*, China University of Geosciences Press (1992).

10.*   Jinfu Deng, Hailing Zhao, Zongxu Wu, and et al. A mantle plume beneath the north part of China continent and lithosphere motion, *Geosci.* 6, 267-274 (1992).

11.*   Jinfu Deng, Hailing Zhao, Delong Ye and et al. Cenozoic volcanic shift and spreading of continental rift and continental drifting, *Experimental Petroleum Geol.* 15, 1-9 (1993).

12.   Jinfu Deng, Laishao Cong, Hailing Zhao and et al. *Petrological record of multiple intracontinental subduction orogency, Himalaya-Gangdise*, INDEPTH Workshop, 16-19 (1994).

13.*   Jinfu Deng, Zongxu Wu, Yianjun Yang and et al. Crust-mantle petrological structure and deep processes along the Golmud-Ejinaqi geoscience section, *Acta Geophys. Sinica*, 38, supp. II, 130-144 (1995).

14.*   Jinfu Deng, Hailing Zhao, Xuanxue Mo, Zongxu Wu and Zhaohua Luo. *Continental roots-plume tectonics of China -Key to the continental dynamics*, Geol. Pub. House (1996).

15.   D.M.Fountain. The Iverea-Verbano and Strona-Ceneri zones northern Italy: A cross-section of the continental crust- new evidence from seismic velocities of rock sample, *Tectonophys.* 33, 145-165 (1976).

16.   D.M.Fountain and M.H. Salisburg. Exposed cross-section through the continental crust: Implication for crustal structure, petrology and evolution, *Earth Planet. Sci. Lett.* 56, 263-277 (1981).

17.   J.R.Holloway and B.J. Wood. *Simulating the Earth, experimental geochemistry*, Unwin Hyman (1988).

18.   H.Kern. P- and S-wave velocities in crustal and mantle rocks under the simultaneous action of high confining pressure ang high temperature and effect of rock microstructure. In: *High-pressure reserches in geoscience*. pp.15-45. Schweif. Verlag (1982).

19.*   Xianjie Shen, Wenren Zhang, Shuzhen Yan and et al. Heat flow and tectono-thermal evolution of terranes of the Qinghai-Tibet plateau. *Geol. Memoirs*, Ser. 5, Numb. 14 (1992).

20.*   Xianjie Shen, Shuzhen Yan and Jiying Shen. Heat flows study and analysis along the Golmud-Ejinaqi geosciences section, *Acta Geophys.Sinica*, 38, supp. II, 86-97 (1995).

21.   S.R.Talor and S.M. Mclennan. *The continental crust: Its composition and evolution*. Blackwell Sci. Pub. (1985).

22.*   Maoji Wang and Zhenyan Chen. Isostatic gravity anomaly and crustal structure, *Acta Geol. Sinica*. 56, 51-61 (1982).

23.   B.L.Weaver and J. Tarney. Empirical approach to estimating the composition of the continental crust, *Nature* 310,575-577 (1984).

24.*   Zhongxu Wu and Chaihua Guo. Petrological model for structure of continental crust in East Hebei province, *Seism. Geol.* 13, 369-376 (1991).

25.*   Zhongxu Wu and Chaihua Guo. High pressure and temperature experimental study of seismic velocities of the crust rock sample, East Hebei province, *Geophys. Develop.* 8, 206-212 (1993).

26.*   Zongxu Wu and Jinfu Deng. 500C isotherm and earthquake of the North China continent, *Geophy. Soc. Ann.* 118 (1994).

27.*   Zongxu Wu, Jinfu Deng, Hailing Zhao and et al. Crust-mantle petrological structure and evolution of the North China continent, *Acta Petrol. Miner.* 13, 106-115 (1994).

28.   P.J.Wyllie. Plate tectonics and magma genesis, *Geol. Rund.* 70, 128-153 (1981).

Proc. 30*th* Int'l. Geol. Congr., Vol. 4, pp. 11-20
Hong Dawei (Ed)
© VSP 1997

The Composition and Ages of Lower Crust along Ningde-Hukou Transect: Constraints from Geochemistry, Neodymium Isotope and Detrital Zircon Ages

FENGQING ZHAO[1], WENSHAN JIN[1], XIAOCHUN GAN[1]
and DAZHONG SUN[2]
[1] *Tianjin Institute of Geology and Mineral Resources, CAGS, Tianjin 300170, P. R. China*
[2] *Guangzhou Institute of Geochemistry, CAS, Guangzhou 510640, P. R. China*

Abstract

The transect runs from Ningde City of Fujian Province to Hukou City of Jiangxi Province and crosses the Huaxia Block in east and Yangtze Block in west bounded by Jiangshao Fault. The oldest basement along the transect is the Paleoproterozoic rocks with the ages of 2400-2000 Ma for Huaxia Block and 2200-2000 Ma for Yangtze Block. These rocks have been metamorphosed to amphibolite facies (Huaxia Block) or high-grade greenschist to low-grade amphibolite facies (Yangtze Block), representing the characteristics of the middle crust along the transect. Most T_{DM} ages and detrital zircon ages in igneous rocks range from 2750 Ma to 2500 Ma and some are up to 3100 Ma, which give an age information of lower crust. A lot of the granitoids in Huaxia Block are S-type ones and have negative ε_{Nd} values, therefore it indicates that the composition of lower crust is composed mainly of felsic protoliths. The ε_{Nd} of Caledonian granitoids ranging from -11 to -8 and ε_{Nd} of Yanshanian granitoids from -6 to -1 infer that the composition of the lower crust in Huaxia Block evolved from granite-rich ("fertile") to granite-poor ("sterile") that were caused either by crustal differentiation or by the addition of mantle components. The granitoids and acid volcanics from Yangtze Block are also peraluminous S-type ones mainly, whereas have ε_{Nd} from -6 to $+4$, which suggest that the lower crust of Yangtze Block have been added by great amount of mantle components and the composition of the lower crust of Huaxia Block is more evolved than Yangtze Block.

Keywords: Ningde-Hukou transect, Composition and ages of lower crust, Huaxia Block, Yangtze Block, T_{DM}, Detrital zircon ages.

INTRODUCTION

Estimating the ages and composition of lower continental crust is a rather difficult task. This is partly due to the complexity of continental crust and poorly understanding of uncertainty over the nature of lower crust. Many attempts have been made to reveal reasonable information of lower crust, such as from seismic reflection and refraction [4,14], from deep crustal xenoliths [12,13] or from exposed deep crustal terrains [1,2]. The composition and structure of lower crust seem to be not static ones, they may have undergone fundamental chages with time. How can we decipher this geodynamical process? Geochemistry, especially isotopic geochemistry, can be used as probe into the deep crust. It can give some information on the composition and the compostional change with time. In this investigation, we report geological, geochemical and geochronological studies on the nature of lower crust along Ningde-Hukou transect.

SEISMIC VELOCITY STRUCTURE OF CRUST ALONG THE TRANSECT

In the recent years, several geophysical transects have been done from south China including

Ningde-Hukou [15], Wenzhou-Tunxi [18], Quanzhou-Heishui [11] transects (Fig. 1). Ningde-Hukou transect runs from the Ningde City of Fujian Province, through Zhenghe-Chong'an-Yongping-Leping, to Hukou City of Jiangxi Province. It crosses the Huaxia Block (Cathaysia) in east and Yangtze Block in west bounded by Jiangshao Fault. According to the seismic velocity structure, the crust along the transect, with total thickness about 30 km, can be divided into upper crust, middle crust and lower crust with the boundary depth at 10 km and at 20 km. The P-velocities(Vp) of upper crust range from about 4.6 km/s to 6.0km/s and the densities from 2.67g/cm³ to 2.75g/cm³, suggesting that the upper crust is composed mainly of sedimentary rocks, granites, volcanics, slates and

Figure 1. Sketch map showing the tectonic framework and the location of geoscience transects from south China, transect 1, 2, 3 representing the Wenzhou-Tunxi, Ningde-Hukou and Quanzhou-Heishui, respectively

phyllites. The middle crust have the Vp from 6.06 km/s to 6.25 km/s and densities from 2.79g/cm³ to 2.85g/cm³, consisting of metamorphic rocks of amphibolites facies. The low-velocity zone (Vp≈5.9km/s) was found in eastern part of Huaxia Block at the depth from 14 km to 17 km, implying the detachment zone in the basal part of the middle crust. The Vp span of lower crust ranging from 6.30km/s to 7.15 km/s overlaps the velocity range of felsic to mafic granulites. The Vp of upper part of lower crust are less than 7.0 km/s, which is corresponding to the velocity intervals of felsic to intermediate granulites [14]. The Vp of basal part of lower crust is more than 7.0 km/s, implying more basic in composition than upper part that may be caused by addition of mantle components.

METAMORPHIC BASEMENT ALONG TRANSECT

The metamorphic rocks of Huaxia Block are divided into three successions: Mayuan Group of Paleoproterozoic, Mamianshan Group of Mesoproterozoic and Changting Supergroup of Neoproterozoic to early Paleozoic (Fig. 2). The Mayuan Group consists of biotite-plagioclase

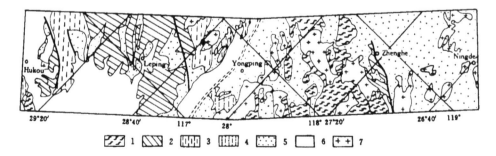

Figure 2. Geological sketch map along Ningde-Hukou transect, showing the distribution and characteristics of metamorphic basement. 1-Paleoproterozoic rocks; 2-Mesoproterozoic rock; 3-Neoproterozoic rocks; 4-Neoproterozoic to early Paleozoic rocks; 5-Cambrian to Cretaceous rocks; 6-Quaternary; 7-Granitoids

gneisses, garnet-biotite-plagioclase gneisses, garnet-mica schists, mica-quartz schists intercalated with amphibolites and marbles. The rocks have been metamorphosed to amphibolite facies with temperatures from 550℃ to 680℃ and pressures from 400MPa to 650MPa [7], inferring burial depths from 16km to 24km. The fluid composition in upper part of Mayuan Group mainly is H_2O, whereas the lower part is rich in CO_2+H_2O. The rocks have been exprienced at least four facies deformation and the former two are recumbent folds of ductile-rheologic mechanism [21]. A lot of single zircon U-Pb istopic ages suggest it is between 2400Ma and 2000Ma [3].

The Mamianshan Group is in structural contact with Mayuan Group, which consists of metamorphic bimodal spilite-keratophyre suite(occur as epidote amphibolites and albite laptites), staurolite-garnet-mica schists, mica-quartz schists, marbles and quartzites. It has been undergone high-grade greenschist to low-grade amphibolite facies metamorphism. Numerous isotopic dating inferred that Mamianshan Group formed between 1400Ma and 1000Ma [3].

The Changting Supergroup is in structural contact with Mayuan Group, which consists of meta-siltstones, feldspar-quartz sandstones, slates, phyllites and graywackes in upper part and plagioclase-mica-quartz schists, quartz-biotite schists in lower part. The rocks have been undergone greenschists facies metamorphism and intensely deformation.

According to above-mentioned characteristics, the rocks of Mayuan Group may contributed to reflection feature of the middle crust and Changting Supergroup can represent the folded basement of upper crust. The Mamianshan Group with locally distribution may exist at transitional zone between upper and middle crust [23].

In the southeastern margin of Yangtze Block, the metamorphic rocks can be divided into the Xingzi Group of Paleoproterozoic and Shuangqiaoshan Group of Mesoproterozoic (Fig. 2). The Xingzi Group, cropped out near the Lushan Mountains, comprises staurolite-garnet-biotite schists, garnet-mica schists, mica-quartz schists and biotite-plagioclase leptites intercalated with amphibolites, quartzites and minor impure marbles. The rocks range in metamorphic grade from high-grade greenschist to low-grade amphibolite facies with the temperatures from 530℃ to 600℃ and pressures from 400MPa to 570MPa [22], implying burial depths from 15km to 20km. It is a representive of middle crust or upper part of middle crust. Single zircon U-Pb dating suggests it formed from 2200Ma to 2000 Ma [22].

The Shuangqiaoshan Group, in the structural contact with Xingzi Group, comprises sericite slates and phyllitic slates, silty-slates, meta-siltstone, meta-graywackes and tuffaceous slates intercalated with spilites and keratophyres in which meta-volcanics show bimodal suite locally. The rocks have been metamorphosed to low-grade greenschist facies and high-grade greenschist facies locally. A lot of isotopic dating suggest it formed during 1700Ma(?) to 1000Ma.

THE AGES OF LOWER CRUST ALONG TRANSECT

Because no granulite facies terrains are exposed along the transect, the ages of Lower crust are traced mainly by Nd model ages (T_{DM}) and detrital zircon ages from the igneous rocks. The zircon is silicate mineral and crystallized mainly from felsic magma, therefore the detrital zircon ages can give some information on the time of felsic protoliths. The T_{DM} ages of mafic rocks can roughly estimate the "early crust formation age" and the T_{DM} of felsic rocks can be interpreted as the crustal residence time of source rocks.

The Ages of Lower Crust of Huaxia Block

Previous geochronological studies on igneous rocks have found some Pre-2. 5Ga ages. Zhu (1985) first published zircon U-Pb concordant intercept age of 2713Ma from Xinqiao granodiorites [26]. After that, numerous U-Pb concordant ages have been reported: 2516Ma from Tanghu granites [9], 2642Ma from Qinghu granites [10] and 3051Ma from Dehua granites [17]. Zhou (1992) reported Sm-Nd isochron ages of 3125 ± 184Ma from amphibolites of Chenchai Group [24]. These ages fall in the time span between 2750Ma and 2500Ma, some are up to 3100Ma, which shed light on age information of lower crust of Huaxia Block. Further isotopic dating by single zircon U-Pb method also yield detrital zircon ages from 2415Ma to 2589Ma(Table 1), supporting the idea that 2750Ma to 2500Ma felsic protoliths underlie Huaxia Block.

Table 1　Isotopic analyses of U-Pb dating for detrital zircons from igneous rocks along transect

U (ppm)	Pb$_{tot.}$ (ppm)	Pb$_{com.}$ (pg)	Atomic ratios[a]					Apparent ages (Ma)		
			$\frac{206Pb}{204Pb}$[b]	$\frac{208Pb}{206Pb}$[c]	$\frac{206Pb}{238U}$[c]	$\frac{207Pb}{235U}$[c]	$\frac{207Pb}{206Pb}$[c]	$\frac{206Pb}{238U}$	$\frac{207Pb}{235U}$	$\frac{207Pb}{206Pb}$
F937, Quartz-Keratophyres of Mesoproterozoic, Fujian Privince, Huaxia Block										
807	248	—	16943	0. 1	0. 2943(30)	5. 784(60)	0. 1431(2)	1670	1940	2258
1418	324	—	4962	0. 3	0. 2138(21)	3. 930(38)	0. 1331(2)	1246	1625	2142
683	410	—	406	0. 1	0. 4167(43)	8. 974(98)	0. 1561(5)	2240	2410	2415
307	158	—	215	0. 1	0. 3691(80)	6. 900(252)	0. 1361(30)	2025	2090	2172
F042, Leptites of Neoproterozoic, Fujian Province, Huaxia Block										
—	—	—	—	—	0. 1951(48)	4. 659(133)	0. 1732(21)	1149	1760	2589
F320, Keratophyres of Neoproterozoic, Fujian Province, Huaxia Block										
712	128	8. 6	3726	0. 1399	0. 1627(7)	2. 700(65)	0. 1203(27)	972	1328	1961
F9305, Yanshanian Granites, Fujian Province, Huaxia Block										
37	31	110	114	0. 0208	0. 4781(51)	10. 788(370)	0. 1636(52)	2519	2505	2494
J522, Amphibolites of Paleoproterozoic, Jiangxi Province, Yangtze Block										
136	337	—	30	0. 4079	0. 5234(199)	13. 91(145)	0. 1927(188)	2714	2743	2765
J9328, Amphibolites of Paleoproterozoic, Jiangxi Province, Yangtze Block										
101	36	11	1831	0. 0736	0. 3260(19)	7. 029(124)	0. 1564(25)	1819	2115	2417
J147, Jinningian Granites, Jiangxi Province, Yangtze Block										
179	116	93	576	0. 0893	0. 5398(20)	14. 245(84)	0. 1914(9)	2782	2766	2754
62	32	11	1441	0. 0392	0. 4737(24)	12. 838(100)	0. 1966(12)	2500	2668	2798
J160, Jinningian Granites, Jiangxi Province, Yangtze Block										
60	80	610	32	0. 0462	0. 2982(130)	4. 320(1006)	0. 1050(234)	1682	1697	1716
J001, Rhyolites of Neoproterozoic, Jiangxi, Yangtze Block										
51	16	13	654	0. 1107	0. 2711(20)	5. 584(149)	0. 1494(37)	1546	1914	2339
43	16	18	341	0. 1275	0. 3006(29)	5. 381(305)	0. 1298(69)	1694	1882	2096
J93113, Tuffaceous Slates of Mesoproterozoic, Jiangxi Province, Yangtze Block										
190	67	2. 4	17332	0. 0758	0. 3348(13)	6. 513(44)	0. 1411(8)	1862	2048	2240
J9367, Keratophyres of Mesoproterozoic, Jiangxi Province, Yangtze Block										
269	99	4. 1	13526	0. 0918	0. 3456(13)	6. 930(31)	0. 1454(4)	1913	2103	2293
J9363, Spilites of Mesoproterozoic, Jiangxi Province, Yangtze Block										
39	12	2. 4	3111	0. 1287	0. 2924(24)	4. 274(194)	0. 1060(45)	1654	1688	1732
J9363, Rhyolites of Neoproterozoic, Jiangxi Province, Yangtze Block										
77	16	21	428	0. 144	0. 1748(13)	2. 429(164)	0. 1008(64)	1039	1251	1638
43	18	39	242	0. 1124	0. 3082(32)	6. 496(256)	0. 1529(56)	1732	2045	2378

[a]All errors are at 2σ levels. [b]Corrected for blank and spike. [c]Corrected for blank, spike and initial Pb.

Table 2 Sm.Nd isotopic composition of granitoids from Huaxia Block

Sample No.	Locality	Sm (ppm)	Nd (ppm)	$\frac{^{147}Sm}{^{144}Nd}$	$\frac{^{143}Nd}{^{144}Nd}$	T_{DM} (Ma)	$f_{Sm/Nd}$	$\varepsilon_{Nd}(t)$	Strat. ages(Ma)	Reference
Zhongtiao Granitoids										
S1-W	Zhejiang	16.50	90.30	0.11054	0.511352	2641	-0.44	-3.1	2000	Wang et al
S2-W	Zhejiang	7.25	38.88	0.11284	0.511374	2668	-0.42	-3.2	2000	(1992)
S3-W	Zhejiang	20.09	116.50	0.1045	0.511275	2603	-0.47	-3.0	2000	
S3-P	Zhejiang	1.09	6.12	0.10727	0.511323	2602	-0.45	-2.6	2000	
S3-G	Zhejiang	7.71	39.81	0.11721	0.511447	2674	-0.40	-3.0	2000	
Calidonian Granitoids										
FZ-2	Fujian	3.640	15.938	0.1381	0.512016	2277	-0.30	-8.8	450	Yuan et al
FZ-6	Fujian	2.593	12.398	0.1265	0.511900	2176	-0.36	-10.4	450	(1991)
F3017	Fujian	5.4181	29.4836	0.11116	0.511269	1400	-0.43	-6.01	450	Huang et al
F3403	Fujian	8.2222	41.891	0.12016	0.51154	1800	-0.40	-9.41	401	(1986,1989)
JNG-001	Jiangxi			0.12604	0.512005	1983	-0.36	-8.5	426	
JW-33150	Jiangxi			0.11944	0.511955	1925	-0.39	-9.2	422	
JLA-1	Jiangxi			0.11208	0.512086	1593	-0.43	-5.9	450	
JYW-75	Jiangxi			0.11879	0.511905	1993	-0.40	-10.4	400	
JCZ-1	Jiangxi			0.11011	0.512048	1618	-0.44	-6.3	450	
J92048	Jiangxi	5.734	27.262	0.1272	0.512098	1848	-0.35	-6.6	450	This paper
J93-22	Fujian	7.978	43.220	0.1116	0.511982	1739	-0.43	-7.9	450	
J-19	Fujian	6.589	33.539	0.1185	0.511998	1800	-0.40	-1.79	535	Huang et al
F-1	Fujian	5.80	32.60	0.1075	0.511423	2500	-0.45	-12.4	470	(1993)
T$_{Nd}$1	Fujian	3.2071	15.623	0.1235	0.512135	1711	-0.37	-5.6	450	
T$_{Nd}$4	Fujian	2.826	14.895	0.1142	0.511692	2224	-0.42	-13.8	450	
Yanshanian Granitoids										
F93-20	Fujian	6.154	36.954	0.1007	0.511814	1787	-0.48	-14.7	120	This paper
F93-5	Fujian	7.744	37.734	0.1241	0.511984	1984	-0.36	-11.8	120	
F-ch-H	Fujian	12.535	39.420	0.19235	0.512344	5679	-0.02	-5.7	120	Huang et al
F-ch-D	Fujian	7.4453	22.765	0.19884	0.512594	5751	-4.8	-0.014	120	(1986)
F-3107	Fujian	2.252	13.069	0.104329	0.512374	1081	-0.46	-3.8	120	
F-Y-H	Fujian	4.7554	23.013	0.1245	0.512325	1408	-0.36	-5.0	120	
F-3156	Fujian	6.7377	39.797	0.10241	0.512378	1057	-0.48	-3.7	120	
F-2h-s	Fujian	5.8988	36.401	0.098024	0.512325	1086	-0.50	-4.7	116	
F-N-sh	Fujian	3.3815	21.908	0.093576	0.512374	985	-0.52	-3.7	116	
F-3153	Fujian	0.6925	6.3854	0.065602	0.512448	723	-0.66	-1.8	116	
BJ-123-82	Taiwan	5.90	28.96	0.1241	0.512390	1292	-0.37	-3.8	116	John et al
BJ-123-82	Taiwan	6.41	31.32	0.1249	0.512402	1282	-0.36	-3.6	116	(1984)
CM22	Fujian	3.80	21.93	0.1046	0.512249	1260	-0.47	-6.3	110	Martin et al
CM24	Fujian	6.64	36.54	0.1098	0.512270	1290	-0.46	-5.7	110	(1990)
CM27	Fujian	6.62	37.62	0.1064	0.512285	1230	-0.46	-5.1	110	
F-3068	Fujian	4.5674	22.716	0.12163	0.512314	1382	-0.38	-5.4	98	Huang et al
F-3367	Fujian	7.111	42.228	0.10902	0.512243	1319	-0.45	-6.6	98	(1986)
F-3067	Fujian	3.1368	18.491	0.10261	0.512280	1193	-0.48	-5.8	99	
Qu-1	Fujian	3.482	15.408	0.136700	0.512458	1368	-0.31	-2.8	91	
F-K-Q	Fujian	5.957	22.717	0.15863	0.512460	1903	-0.19	-3.1	91	
F-3371	Fujian	8.171	30.647	0.16192	0.512474	1982	-0.17	-2.8	91	
F-K-H	Fujian	3.3762	20.115	0.10153	0.512482	908	-0.48	-2.0	91	
F-JD	Fujian	5.1265	35.270	0.08775	0.512433	868	-0.56	-2.8	91	
CM-11	Fujian	10.28	41.19	0.1509	0.512482	1620	-0.23	-2.6	90	Martin et al
CH-14	Fujian	5.36	20.65	0.1568	0.512560	1740	-0.20	-2.2	90	(1990)

The granitoids are widespread in the Huaxia Block, which emplaced from Paleoproterozoic to Mesozoic and show elemental and isotopic characteristics of S-type ones. Therefore granitoids may originate from anatexis of felsic protoliths, which can be used as remote samples to study the time and composition of lower crust.

The Zhongtiaonian granitoids (1. 9 ± 0. 1Ga) have the T_{DM} ages from 2602Ma to 2674Ma (Table 2 and Fig. 3). They are characterized by low variation of $f_{Sm/Nd}$ and ε_{Nd}[16], which indicates that the Sm and Nd isotopic composition have not been strongly influenced by AFC processes. Therefore, the 2600Ma to 2700Ma can be referred to ages of source rocks.

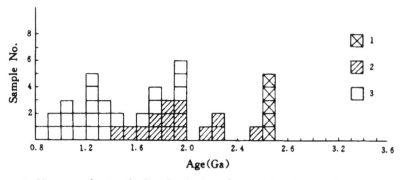

Figure 3. Histogram showing the T_{DM} distributions of granitoids in Huaxia Block. 1-Zhongtiaonian granitoids; 2-Caledonian granitoids; 3-Yanshanian granitoids.

The T_{DM} ages of Caledonian granitoids(0. 4~0. 45Ga) range mainly from 1800Ma to 2500Ma (Table 2 and Fig. 3), which is corresponding to the time span of Mayuan Group. Field evidences show that the granitoids intruded into the rocks of Mayuan Group, so the granitoids should be derieved by partial melting of Pre-2. 5Ga rocks instead of Paleoproterozoic rocks.

The T_{DM} ages of Yanshanian granitoids(0. 1~0. 12Ga) are highly variable and most of them are smaller than 2000Ma, the granitoids have high ε_{Nd} values(Table 2 and Fig. 3), implying that the source rocks have been added by variable great amount of mantle components. Inputing of mantle materials may be the reason of reducing the resulting T_{DM} ages.

The T_{DM} ages from metamorphic basic volcanics (amphibolites) are shown on Fig. 4. The amphibolites of Mayuan Group have the T_{DM} ages from 2400 to 2600Ma and 2200-2300Ma. The Mamianshan Group have the T_{DM} ages mainly from 2000 to 2300Ma. It implies that early underplating events may occurred at the time intervals 2000 to 2300Ma and 2400-2600Ma.

The Ages of Lower Crust of Yangtze Block

A lot of single zircon U-Pb dating has yielded the age message of Pre-2. 2Ga felsic basement underlie the Yangtze block along the transect(Table 1). The ages distributed in 2700 — 2800Ma and 2200—2450Ma. It can be further confirmed by T_{DM} ages of felsic rocks(Table 3).

The T_{DM} ages of mafic volcanics show variable feature but mainly range from 2000-2400Ma and 1300-1700Ma, some are up to 2600-2800Ma(Table 3), which give the age information of underplating events.

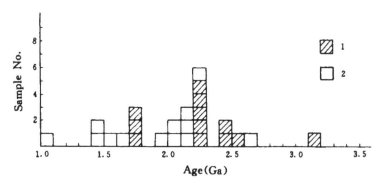

Figure 4. Histogram showing the T_{DM} distributions of amphibolites in Huaxia Block. 1-Paleoprotero-zoic; 2-Mesoproterozoic. Isotopic data are from Zhao et al[23].

Table 3   Sm,Nd isotopic analyses of igneous rocks from Jiangxi Province,Yangtze Block

Sample No.	Sm (ppm)	Nd (ppm)	$\frac{^{147}Sm}{^{144}Nd}$	$\frac{^{143}Nd}{^{144}Nd}$	T_{DM} (Ma)	$f_{Sm/Nd}$	$\varepsilon_{Nd}(t)$	Strat. ages (Ma)
Amphibolites of Paleoproterozoic								
J92137	2.669	8.463	0.1907	0.512741	2695	−0.03	+3.7	2180
J92139	3.779	12.083	0.1891	0.512794	2197	−0.04	+5.2	2180
522−3T[a]	2.957	8.283	0.2158	0.512715	−	+0.10	+3.7	2180
522−1P[b]	1.220	3.768	0.1957	0.512930	1858	−0.01	+6.0	2180
522−2H[c]	2.684	5.844	0.2777	0.512848	−	+0.41	+18.7	2180
Meta-basic Volcanics of Mesoproterozoic								
J92046	4.759	17.974	0.1600	0.512374	2194	−0.19	+0.5	1200
J92069	3.076	11.695	0.1590	0.512214	2594	−0.19	−2.5	1200
J92131	5.687	23.393	0.1470	0.511903	2832	−0.25	−6.8	1183
Basic Volcanics of Neoproterozoic								
J92054	9.614	41.796	0.1391	0.512160	2016	−0.29	−2.7	900
J92061	15.138	69.842	0.1310	0.512394	1391	−0.33	+2.8	900
J92063	13.140	62.379	0.1344	0.512076	2057	−0.32	−3.8	900
J92114	3.140	18.795	0.1010	0.511375	2389	−0.49	−13.7	900
Rhyolites of Mesoproterozoic								
J92158	16.515	76.883	0.1299	0.512271	1596	−0.34	+3.0	1183
J92071	2.463	10.327	0.1442	0.511705	3147	−0.26	−10.2	1200
Rhyolites of Neoproterozoic								
C−12	9.756	44.587	0.1323	0.512405	1393	−0.32	+2.2	817
C−5−2	10.898	49.862	0.1321	0.512164	1873	−0.33	−2.5	817
J92062	17.038	71.017	0.1450	0.512384	1695	−0.26	+0.5	817
J92001	2.545	17.098	0.1488	0.511977	2735	−0.24	−7.8	817
J92005	4.868	28.832	0.1021	0.512299	1161	−0.48	+3.2	817
Jinningian Granitoids								
J92148	3.505	14.366	0.1475	0.512037	2549	−0.25	−6.3	866
J92159	5.343	26.127	0.1236	0.512162	1668	−0.37	−0.6	926

[a]whole rock, [b]Plagioclase, [c]Hornblende

COMPOSITION OF LOWER CRUST ALONG TRANSECT

The elemental and isotopic geochemistry of igneous rocks reveal that the composition of lower crust are very complicated, but consist mainly of felsic protoliths. Major differences exist between two blocks, at different crustal levels and during different time.

The Composition of Lower Crust of Huaxia Block

The Zhongtiaonian granitoids consist of granodiorites, monzogranites and potassium feldspar granites in which most of them are S-type ones with minor I-type or transitional type between I- and S-types. The granitoids have ϵ_{Nd} values from -3.1 to -2.6 and $^{147}Sm/^{144}Nd$ from 0.10 to 0.14, implying important contributions of old felsic components to the source rocks.

The Caledonian granitoids of Huaxia Block include potassium feldspar granites, alkali granites and granodiorites with the S-type feature. The granitoids have negative ϵ_{Nd} values (-8 to -11) and low $^{147}Sm/^{144}Nd$ ratios (0.10 to 0.14), inferring that they are the anatectic products of fertile (granite-rich) sources.

The Yanshanian granitoids are mainly alkali granites, potassium feldspar granites and monzogranites including some alkali-rich granites. Most of them are chiefly S-type ones but the ratios of A-type and I-type to S-type are higher than the former two ages of granitoids and show relatively high ϵ_{Nd} values (-6 to -1). The $^{206}Pb/^{204}Pb$, $^{207}Pb/^{204}Pb$ and $^{208}Pb/^{204}Pb$ ratios of alkali-rich granites are similar to other types of Yanshanian granites [20], which inferred that the Yanshanian granitoids were formed by partial melting of ancient crustal rocks in which alkali-rich granites involved great amount of mantle components to its source rocks.

Based on the ϵ_{Nd} evolution line (Fig. 5), the compositions of lower crust at Zhongtiaonian are more evolved than Caledonian, which are in turn more evolved than Yanshanian. The ϵ_{Nd} values of Caledonian granitoids ruled out intensely addition of mantle components, so the compositions of lower crust from Zhongtiaonian to Caledonian was caused by the crustal differentiation. The Yanshanian granitoids with relatively high ϵ_{Nd} values could be formed by mixture of crust and mantle components, implying that the composition chang from Caledonian to Yanshanian could due to the addition of mantle materials.

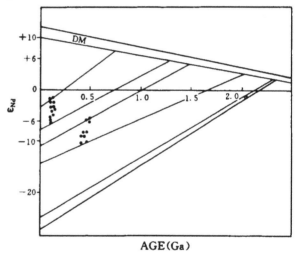

Figure 5. ϵ_{Nd} vs. age diagram for granitoids in Huaxia Block, illustrating the ϵ_{Nd} evolution at different time.

The Compositions of Lower Crust of Yangtze Block

The Jinningian granitoids(0.9±0.05Ga) show K_2O-rich calcalkaline series. It include monzogranites, granites, granodiorites and potassium feldspar granites in which exist some peraluminous minerals like cordierite, garnet and biotite and peraluminous enclaves of silliminate-mica schists[25]. Geochemical evidences also exhibit S-type feature. On the other hand, the granitoids have relatively high ϵ_{Nd} values (-6 to $+4$)(Fig. 6) and low initial $^{87}Sr/^{86}Sr$ ratios [25], implying variable great amount of mantle components to the source rocks of granitoids. In addition, some M-type granites are exposed along the contact zone between Huaxia and Yangtze Blocks, which also show that the compositions of lower crust of Yangtze Block are more depleted in felsic components than Huaxia Block.

The rhyolites of Neoprotero-zoic (0. 8-0. 9Ga) have mineralogical and chemical characteristics of S-type ones. In contrast, the ε_{Nd} of rhyolites range from -1.9 to $+2.8$(Fig. 6), showing the similar geochemical character to the Jiningian granites.

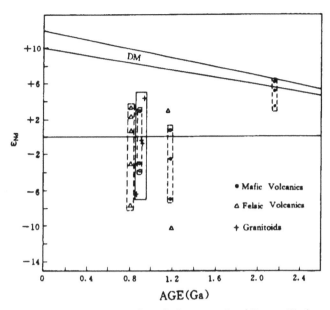

Figure 6. ε_{Nd} vs. age diagram from the igneous rocks of Yangtze Block.

CONCLUSION

The geological, geochemical and geochronological characteristics described above provide some reasonable constraints on the nature of lower crust along Ningde-Hukou transect, some general conclusion can be drawn as follow:

(1) Elemental and isotopic evidences from the granitoids and acid volcanics show that the composition of lower crust is composed mainly of felsic protoliths especially during the early stage of crust-forming process. It is further confirmed by geophysical approaches. The seismic structure of Ningde-Hukou and other transects from south China indicate that the Vp of lower crust are chiefly less than 7. 0 km/s except the basal part, which is corresponding to Vp range from acid to intermediate components.

(2) Major differences of compositions of lower crust exist between the two Block. The Yangtze Block exhibit less evolved feature in composition than Huaxia Block, which is due to more intensely addition of mantle components to lower crust of Yangtze Block.

(3) The evidences from isotopic geochemistry demonstrate that the lower crustal compositions have significant changes with time. The evolution trend is from granite-rich to granite-poor, which is caused either by crustal differentiation or addition of mantle materials.

(4) Detail studies on the T_{DM} ages and detrital zircon ages from igneous rocks indicate that the lower crust along the transect may have been formed from 2750 Ma to 2500 Ma, some is up to 3100 to 3000 Ma. After that, episodic underplating of mantle perhaps occurred in the time intervals 2400 – 2500 Ma, and 2200 – 2300 Ma in Huaxia Block and 2000-2400 Ma, and 1300-1600 Ma in Yangtze Blocks.

REFERENCES

1. D. M. Fountain and M. H. Salisbury. Exposed cross-sections through the continental crust; Implications for crustal structure, petrology and evolution. *Earth Planet. Sci. Lett.* **90**, 263-277(1981).
2. D. M. Fountain, R. Arculus and R. W. Kay. Continental lower crust. *Elsevier*, Amsterdam, (1992).

3. Gan Xiaochun, Li HM, Sun DZ, Zhuang JM. Geochronological study on the Precambrian metamorphic basement in northern Fujian. *Geology of Fujian.* 12, 17-32(1993).

4. W. S. Holbrook, W. D. Mooney and N. I. Christensen. The seismic velocity structure of the deep crust. In: *Continental lower crust.* D. M. Fountain, R. Arculus and R. W. kay (Eds). pp1-43. Elsevier, (1992).

5. Huang Xuan, Sun SH, D. J. Depaolo and Wu KL. Nd-Sr study on Cretaceous magmatic rocks from Fujian Province. *Acta Petrologica Sinica.* 2, 50-63(1986).

6. Huang Xuan and D. J. Depaolo. Study of sources of Paleozoic granitoids and the basement of south China by means of Nd-Sr isotope. *Acta Petrologica Sinica.* 5,28-36(1989).

7. Jing Wenshan et al. Characteristics of petrology, geochemistry and metamorphism of the Precaledonian regional metamorphic rocks in Fujian Province. *Geology of Fujian.* 11, 241-262(1992).

8. B. M. John, X. H. Zhou and J. L Li, Formation and tectonic evolution of southeastern China and Taiwan: Isotopic and geochemical constraits. *Tectonophysics.* 183, 145-160(1990).

9. Li Xianhua, M. Tatsumoto and Gei XT. The discovery of 2. 5Ga-old Archean relict zircon from the Tanghu granite, south China, and its preliminary implication. *Chinese Science Bulletin.* 34, 1364-1369(1989).

10. H. Martin, B. Bonin, R. Capdevila et al. The Fuzhou granitic complex(SE China), petrology and geochemistry. *Geochemica.* 2, 101-111(1991).

11. Qing Baohu. The infrastructure of Hunan revealed by Taiwan-Heishui geoscience transect. *Hunan Geology.* 10, 89-96(1991).

12. R. L. Rudnick and S. R. Taylor. The composition and petrogenesis of the lower crust: a xenolith study. *J. Geophys. Res.* 92, 13981-14005(1987).

13. R. L. Rudnick. Xenoliths-samples of the lower continental crust. In: *continental lower crust.* D. M. Fountain. R. Arculus and R. W. Kay(Eds). pp269-316. Elsevier, (1992).

14. S. B. Smithson. A physical model of the lower crust from north America based on seismic reflection data, In: *The Nature of the Lower Continental Crust.* J. B Dawson et al (Eds). pp. 23-24. Geological Society Special Pub. 24, (1986).

15. Wang Chunyong, Lin ZY. and Chen XB. Comprehensive study of geophysics on geoscience transect from Menyuan, Qinghai Province, to Ningde, Fujian Province, China. *Acta Geophysica Sinica.* 38, 590-598(1995).

16. Wang Yinxi, Yang JD, Guo LZ et al. The discovery of the lower Proterozoic granite in Longquan, Zhejiang Province, and the age of the basement, *Geological Review.* 38, 525-531(1992).

17. Wang Zhenmin. New recognition on Cathaysia and its related geological issues. In: *Contribution to Geology of Fujian Province.* pp116-136. Fujian Map Pub. House(1996).

18. Xiong Shaobai, Lai MH, Liu HB, Yu GS. Lithospheric structure and seismic velocity distribution along Tunxi-Wenzhou Transect. In: *Lithospheric structure and Geological Evolution of Southeastern China continent.* Li Jiliang (Ed). pp. 250-256. Metallurgical Industry Pub. (1993).

19. Yuan Zhongxin. and Zhang ZQ. Sm-Nd isotopic characteristics of granitoids in the Nanling Region and their petro-genetic analysis. *Geological Review.* 38, 1-15(1992).

20. Zhang Ligang. Characteristic model of Pb isotopic compositions of alkali-rich igneous rocks from eastern china and its geological significance. *Earth Science-J. China University of Geoscience.* 19, 227-234(1994).

21. Zhao Fengqing, Chen YZ, Li RA. Study on multiple deformation and structured level of Precaledonian metamorphic basement. *Geology of Fujian.* 12, 33-40(1993).

22. Zhao Fengqing, Jin WS and Gan XC. Characteristics and tectonic settings of the Paleoproterozoic metamorphic basement on both sides of the Jiangshao Fault. *Geology of Anhui.* 4, 73-81(1994).

23. Zhao Fengqing, Jin WS, Gan XC, Sun DZ, Wang ZW. Discussion on the characteristics of the deep crust and Precaledonian metamorphic basement in the Cathaysia Block, Southeastern China. *Acta. Geoscientia Sinica.* No. 3, 235-245(1995).

24. Zhou Xinhua, Cheng H. and Zhong ZH. The sources province of metamorphic rock from southern Zhejiang-a ease study. In *"Memoir of Lithosphere Tectonic Evolation Research(1)".* pp. 114-120, Seismology press. (1993).

25. Zhou Xinmin and Wang D. The peraluminous granodiorites with low initial $^{87}Sr/^{86}Sr$ ratio and their genesis in southern Anhui Province, eastern China. *Acta Petrologica Sinica.* No. 8, 37-44(1988).

26. Zhu Yulin. On the age of Xinqiao Massif. *Regional Geology of China.* No. 4, 353-357(1988).

Proc. 30ᵗʰ Int'l. Geol. Congr., Vol. 4, pp. 21-30
Hong Dawei (Ed)

Measurement of P-Wave Velocities in Rocks of Qinling and North China: Its Applications for the Crustal Low Velocity Layers

ZHI-DAN ZHAO[1] SHAN GAO[2] TING-CHUAN LUO[2] BEN-REN ZHANG[2]
HONG-SEN XIE[1] JIE GUO[1] AND ZU-MING XU[1]

1.*Laboratory of Materials of the Earth's Interior, Institute of Geochemistry,Chinese Academy of Sciences, Guiyang, 550002, China*
2.*Institute of Geochemistry, China University of Geosciences, Wuhan, 430074, China*

Abstract

The crustal low velocity layers (CLVL) occuring in the middle-lower crust have been found in many regions around the world and have been explained in different ways. In this paper, the compressional wave velocities (V_p) of 138 rock samples selected from Qinling and North China were measured under high temperatures (up to 1500 °C) and high pressures (up to 3 GPa) simultaneously. The general feature of the experimental data were discussed and the Vp decrease phenomenon (54 samples involved) were found. Detailed studies, mineral constitutions and textures of experimental products examined by microscope and EPMA, indicated that the dehydration, phase transition and partial melt of water-bearing minerals (hornblende and biotite), are responsible for the V_p decrease. The results suggest that dehydration melting of hydrous minerals might be an important mechanism for CLVL observed in the continental crust of the study area and probably other parts of the world.

Key words: crustal lower-velocity layers, compressional velocity in rocks, dehydration and phase transition, partial melt, The Qinling Orogenic Belt, North China Craton.

INTRODUCTION

Since 1950's, the deep crustal survey dominated by geophysics and the experimental research of the materials of the earth's interiors lead us know much more about the lithosphere, especially continental lithosphere. The crustal low velocity layer (CLVL) is one of the most important discoveries found by deep geophysical survey. The CLVL are scatting in many regions, such as Alps, North American, Tibet, North China and the Qinling Orogenic Belt. Most of the areas are corresponding to tectoniclly or seismically active belts with high heat flow [2, 12, 13, 14, 19].

The origin of CLVL has been explained in different ways. Some scientists thought that the CLVL maybe soft layers that caused by liquids, large-scale nappe structure or ductile shear belt [8, 9]. While others study the constitution and physical-chemical processes of crustal minerals and rocks in laboratory, and argue that the layers are possibly caused by graphite, saline fluids, dehydration, phase transition or partial melts [8, 14, 19]. But all the study results update imply that there still no one hypothesis could be the only candidate that causes the CLVL. In our study, the compressional velocities (V_p) in 138 rock samples collected from the Qinling Orogenic Belt and North China Craton were measured under high pressures and high temperatures. The experimental data were dicussed and applicated to explain the origin of CLVL in the study areas.

SAMPLES AND GEOLOGIC BACKGROUND

The study areas include the Qinling Orogenic Belt and North China Craton (Wutaishan and Inner Mongolia). The geologic units that sampled systematically are mainly the Precambrian basements and intrusions of various ages, which are known to be the representative rocks of the the crust (Table 1). The samples include eclogite facies, granulite facies, amphibolite facies and greenschist facies rocks. The specimens, which fresh, free of visible fractures and secondary alteration veins were selected in the experiments. Microscope observation, density measurement and major element chemical analyses were finished on each of the experimental samples. Microscope and EPMA (Electron probe micro-analyser) were also used on the products.

Table 1. Sample sources of the V_p measurements

Tectonic units	Rock units
The Qinling Orogenic Belt and its adjacent regions	NC: Taihua Gr., Dengfeng Gr., Angou Gr., Granites
	NQ: Qinling Gr., Kuanping Gr., Erlangping Gr.,Granites
	SQ: Wudang Gr., Bikou Gr. Yunxi Gr., Yaolinghe Gr., Yudongzi Gr., Foping Gr., Granites
	YC: Kongling Gr., Houhe Gr., Granites
	T-D: Dahe Gr., Dabie Gr., Eclogites
North China Craton (Wutaishan and Inner Mongolia)	Jining Gr., Fuping Gr., Longquanguan Com., Hengshan Com., Wutai Gr., Hutuo Gr., Erdaowa Gr., Majiadian Gr., Wulashan Gr., Seertengshan Gr., Granites

NC, The south margin of North China Craton; NQ, North Qinling Belt; SQ, South Qinling Belt; YC, The north margin of Yangtze Craton; T-D, The Tongbo-Dabie Orogenic Belt; Gr, Group

EXPERIMENTAL METHODS

The experiments were performed in a high P-T cell in the so-called YJ-3000 ton press (maxium force: 3000 ton) at the Laboratory of Materials of the Earth's Interior, Institute of Geochemistry, Chinese Academy of Sciences, Guiyang, China. A sketh map of the experimental system and the sample assembly is as Fig. 1 [18].

The press can contribute P-T conditions inside the pressure cell insures stability of both P and T up to 6.0 GPa and 1600 ℃ in a periods from several minutes to 100h. One sample core 12 mm in diameter and 33 in lenth were cut from each specimen. Pyrophyllite cube, which formed from pyrophyllite powder, are used as pressure-transmitting material. The sample along with heating elements (2 or 3 stainless steel foils) was fitted into the hole drilled through the cube. The temperature of the cell could be attained from the power-temperature experimental curves. The P-wave velocities in the samples under high P-T were calculated from the mumberical signals on the SD-1 sound wave detector meter. A more detailed describe of the sound velocity measurement and calculation method were in Xie [17] and Xu [18].

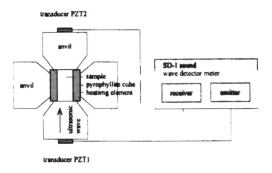

Figure 1. Sample assembly [18].

In order to finely imitate the actural P-T conditions of the Earth's deep crust and upper mantle of the areas, V_P in rocks were measured under high temperature (up to 1500 ℃) and pressure (up to 3.0 GPa) simultaneously according to the geotherm for the regions [7]. The heat elements of stainless foils would be melt into half when the temperature is beyond the melting point of steel. Therefore the melting time is the end of measurement of one sample. Although the end P and T were different , they are all over 1000 ℃ and 1GPa. A large number of data of V_P were measured and were expressed in V_P-Depth relations, in which Depth stand for depth of the lithospere calculated from the geo-pressure gradient 0.03 GPa/km.

GENENERAL CHARACTERISTICS OF V_P

The relations of six of all the samples were chosen in Fig. 2. The relation can be divided into three parts. V_P increase rapidly with depth in the range between 0 and 10 km in the first part, reflecting the closure of cracks in the samples. V_P slowly increase in the range between 10 and 30 km, and reach the maximum at 750-920 ℃ and 0.63-0.90 GPa (corresponding to 21-30 km) in the second part. The features of the samples during the second part (dV_P/dDepth and V_{Pmax}) stand for the intrisic nature of various types of rocks. The third part of the curves is refer to the state after V_{Pmax}. Two trends are showing, one is the curves nearly content close to the V_{Pmax} with depth increasing, and the other is V_P gradually decreases to the minimum at 1100-1200 ℃ and 0.99-1.50 GPa (30-50 km). The so-called V_P decrease phenomenon showing in 54 samples (more than 1/3 of all the 138 samples), provide key evidence for the origin of CLVL.

FEATURES OF THE SAMPLES WITH V_P DECREASE PHENOMENON

V_P features of the 54 rock samples were summarized in Table 2, five of them were showing in Fig. 3. The products of all the 138 samples were also observed carefully in

24

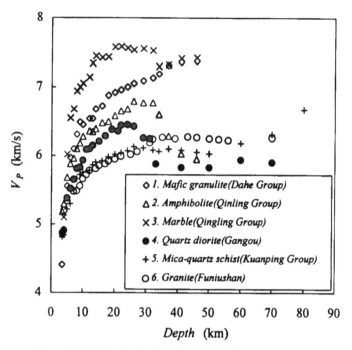

Figure 2. V_P-Depth relations for some of the experimental samples.

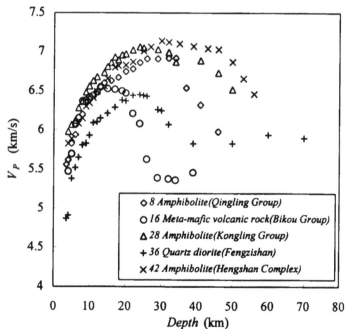

Figure 3. V_P-Depth relations for some of the samples showing the Vp-decrease phenomenon (Sample number are the same as Tab. 2).

Table 2. Measurements of samples with V_p decrease phenomenon in the Qinling Orogenic Belt and North China Craton.

No.	Rock Unit	Rock	V_{Pmax} km/s	P GPa	T ℃	V_{Pmin} km/s	P GPa	T ℃	ΔV km/s	ΔV %
1	Taihua Gr.	Amphibolite	7.39	0.75	650	6.41	2.10	1300	0.98	13.26
2	Dengfeng Gr.	Gabbro	7.66	0.90	750	7.04	1.59	1110	0.62	8.09
3	Angou Gr.	Meta-sed. rock	7.16	0.60	550	6.16	1.59	1110	1.00	13.97
4	Angou Gr.	Meta-basic volc. rock	6.89	0.81	700	6.18	1.50	1060	0.71	10.30
5	Qinling Gr.	Marble	7.55	0.87	920	5.95	1.50	1230	1.60	21.19
6	Qinling Gr.	Amphibolite	7.20	1.02	1040	5.83	1.23	1160	1.37	19.03
7	Qinling Gr.	Amphibolite	6.97	0.78	860	6.57	1.02	1040	0.40	5.74
8	Qinling Gr.	Amphibolite	6.92	1.02	1040	5.98	1.38	1160	0.94	13.58
9	Qinling Gr.	Amphibolite	6.77	0.96	980	5.95	1.38	1230	0.82	12.11
10	Qinling Gr.	Marble	6.35	0.72	800	5.83	1.02	1040	0.52	8.19
11	Kuanping Gr.	Amphibolite	6.53	0.72	800	6.08	1.38	1230	0.45	6.89
12	Erlangping Gr.	Meta-basic volc. rock	6.66	0.87	920	5.91	1.38	1160	0.75	11.26
13	Wudang Gr.	Meta-clastic rock	6.28	0.72	860	5.76	1.26	1230	0.52	8.28
14	Wudang Gr.	Meta-basic volc. rock	6.54	0.63	750	5.50	1.14	1230	1.04	15.90
15	Wudang Gr.	Meta-felsic volc. rock	6.50	0.78	920	5.63	1.02	1160	0.87	13.38
16	Bikou Gr.	Meta-basic volc. rock	6.50	0.63	750	5.36	1.02	1160	1.14	17.54
17	Yunxi Gr.	Meta-basic volc. rock	5.84	0.63	750	5.60	1.02	1160	0.24	4.11
18	Yaolinghe Gr.	Meta-basic volc. rock	6.15	0.63	750	5.70	1.02	1160	0.45	7.32
19	Yaolinghe Gr.	Meta-felsic volc. rock	6.26	0.93	1040	6.08	1.02	1160	0.18	2.88
20	Yaolinghe Gr.	Mica-pl schist	6.65	0.66	800	5.69	1.26	1300	0.96	14.44
21	Yudongzi Gr.	Amphibolite	6.00	0.63	750	5.59	1.02	1160	0.41	6.83
22	Yudongzi Gr.	Felsic gneiss	5.93	0.63	750	5.61	1.02	1160	0.32	5.40
23	Foping Gr.	Granulite	6.62	0.57	700	5.92	1.02	1160	0.70	10.57
24	Foping Gr.	Granulite	6.87	0.36	500	5.70	0.93	1040	1.17	17.03
25	Shuangxi Z	Garnet gneiss	5.95	0.63	750	5.60	0.99	1100	0.35	5.88
26	Kongling Gr.	Amphibolite	6.80	0.61	700	6.09	1.50	1230	0.71	10.44
27	Kongling Gr.	Amphibolite	7.05	0.66	750	6.51	1.35	1160	0.54	7.66
28	Kongling Gr.	Amphibolite	7.06	0.72	800	6.31	1.50	1230	0.54	7.65
29	Houhe Gr.	Bitite gneiss	6.20	0.72	800	5.95	1.02	1040	0.25	4.03
30	Houhe Gr.	Biotite gneiss	7.07	0.66	750	5.94	1.02	1040	1.13	15.98
31	Dahe Gr.	Felsic granulite	6.19	1.02	1040	5.92	1.50	1300	0.27	4.36
32	Dabie Gr.	Marble	7.11	0.66	750	6.69	1.38	1160	0.42	5.91
33	Hong'an Gr.	Garnet amphibolite	7.11	0.84	920	6.75	1.71	1300	0.36	5.06
34	Banshanping	Quartz diorite	6.36	0.72	800	6.10	1.23	1160	0.26	4.09
35	Gangou	Quartz diorite	6.44	0.66	800	5.83	1.20	1100	0.61	9.47
36	Fengzishan	Quartz diorite	6.00	0.63	750	5.68	0.99	1100	0.32	5.33
37	Bijigou	Gabbro	6.88	0.57	700	5.43	1.02	1160	1.45	21.08
38	Xichahe	Diorite	6.25	0.63	750	5.92	1.02	1160	0.33	5.28
39	Sandouping	Quartz diorite	6.49	0.96	980	6.22	1.20	1060	0.27	4.16
40	Sandouping	Quartz diorite	6.22	0.96	980	6.12	1.35	1110	0.10	1.61
41	Hengshan Com.	Amphibolite	7.14	0.90	750	6.46	1.68	1230	0.68	9.52
42	Hengshan Com.	Amphibolite	6.96	1.38	1040	6.28	1.68	1230	0.68	9.77
43	Hengshan Com.	Granulite	7.07	1.38	1040	6.83	1.59	1160	0.24	3.39
44	Hengshan Com.	Granulite	7.20	1.17	920	6.56	1.92	1230	0.64	8.89
45	Wutai Gr.	Basic volc. rock	6.39	0.90	750	6.06	1.29	980	0.33	5.16
46	Wutai Gr.	Felsic volc. rock	6.29	0.90	750	6.17	1.68	1230	0.12	1.90

Table 2. (continued)

No.	Rock Unit	Rock	V_{Pmax} km/s	P GPa	T $°C$	V_{Pmin} km/s	P GPa	T $°C$	ΔV km/s	ΔV %
47	Wutai Gr.	Felsic volc. rock	7.77	1.17	920	6.80	1.68	1230	0.97	12.48
48	Fuping Gr.	Pyroxenite	7.03	0.90	750	6.38	1.92	1300	0.65	9.25
49	Hutuo Gr.	Basic volc. rock	6.49	0.75	650	6.02	1.38	1040	0.47	7.24
50	Jining Gr.	Granulite	7.31	1.29	980	7.05	1.68	1230	0.26	3.56
51	Erdaowa Gr.	Biotite schist	7.70	1.05	860	7.24	1.68	1230	0.46	5.97
52	Majiadian Gr.	Actinolite slate	6.85	0.75	650	5.88	1.38	1040	0.97	14.16
53	Wulashan Gr.	Amphibolite	7.10	0.90	750	6.26	1.59	1160	0.84	11.83
54	Wulashan Gr.	Pyroxene granulite	6.97	1.05	860	6.71	1.68	1230	0.26	3.73

1. Gr., Group; Com.., Complex; sed., sedimentary; volc., volcanic. 2. V_{Pmax} and V_{Pmin} are referred to the maximum and maximum V_p respectively. P and T are the pressures and temperatures corresponding to V_{Pmax} and V_{Pmin}. 3. $\Delta V_p = V_{Pmax} - V_{Pmin}$. 4. $\Delta V_p (\%) = (\Delta V_p / V_{Pmax}) \times 100\%$.

order to find the mechanism that caused the V_P decrease phenomenon. Most of the products of the experiment have been observed by microscope and checked by EPMA in detail.

The 54 rocks are mainly amphibolite, quartz-diorite, granulite, gabbro, marble, biotite gneiss. Except for marbles, nearly all the 54 samples containing hornblende or biotite. If we check all the 138 samples, we can find that the rocks containing hornblende were all having V_P decrease in their V_P-H relations.

Most of the samples reach the maximum $V_P (V_{Pmax})$ in the ranges of 750-920 $°C$ and 0.63-0.90 GPa (corresponding to 21-30 km), and decrease to the minimum $V_P (V_{Pmin})$ in the ranges of 1100-1200 $°C$ and 0.99-1.50 GPa (corresponding to 21-30 km). Surposing $\Delta V_P = V_{Pmax} - V_{Pmin}$, ΔV_P of seven samples are more than 1.0 km/s, and they are 2 amphibolites, 2 metamophic basic volcanic rocks, 1 gabbro, 1 granulite and 1 marble. The maximum decrease of V_P is 21% in No. 5 and No. 37 (Table 2). The features of three major types of samples are described below.

Rocks bearing hornblende or biotite
All the 138 experimental rocks that bearing hornblende or biotite have V_P decrease. We can see piece of glasses (stand for melting) in the margins of hornblende or biotite (for instance in No.18). Melting glasses are appearing in the boundary between hornblende (or Biotite) and light color minerals (Plaseclase, Quartz), which are colorless, brown, light-yellow. Melt glasses could be about 5-10% of bulk volume of the sample. The results of EPMA of the samples confirmed that some horblende has changed into pyroxene by dehydration and phase transition (i.e., in No. 39) [20].

Marble
V_P decrease phenomenon has been found in all the marbles in our experiments. Marbles changed from light-color and colorlessness of the started samples to green or dark-green of the products. Some small holes (less than 1 mm) appeared, which

possibly the structure after the escape of CO_2 of the carbonates (i.e., calcite or dolomite). A black material, possibly standing for the products of carbonate's phase transition, can be seen under microscope [20].

Granite
Some V_P-Depth relations of granites were shown in Fig. 4, in which V_P increase with increasing depth but does not decrease. The granites do not bear horblende or bear little biotite. The products display little change of minerals except for developing some microcracks. No melts were found in the samples. We could be sure that dehydration, transition or partial melting has not occurred in granites, mainly because there is not water-bearing minerals.

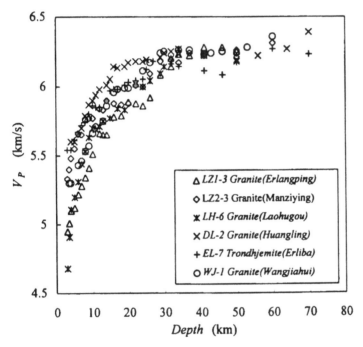

Fig. 4 V_P -Depth relations for some the Granites.

We could get a short summary base on the results above. The digassing of carbonates and phase transition caused V_P decrease in marbles, and left some small holes. The V_P decrease for the rocks containing water-bearing minerals (horblende or biotite) are mainly caused by dehydration, phase transition and partial melting.

DISCUSSION

Lots of experimental results of elastic velocities in rocks indicated that, V_P increase simply with increasing pressures at room temperature, and decrease with increasing temperatures under constant pressures. It is undoubtedly that temperature is the basic reason that caused V_p decreasing in rocks [1, 4, 11].

Many researchers have explained the mechanics why temperature cause V_p decrease. Kern [10] believed that the opening of microcracks could happen less than 0.6GPa. Christensen [1] considered that the effects of grain boundary porosity on rock velocities above 1GPa is minimal, that means microcracks is not an important factor under the conditions equal to the deep crust. Other results support the views that dehydration and phase-transition, such as α-β transition in Quartz [10], could cause V_p decreasing. We prefer to that, based on our experimental data, dehydration, phase-transition and partial melting happened in the rocks bearing horblende or biotite, and caused V_p decreasing   of the whole samples.

We try to reach the actual conditions of deep crust in our experiments, namely, both temperatures and pressures increase during the measurements. Large number of boundary cracks are very difficult to open. The 138 rock samples displayed the same feature that they all developed cracks, not only the 54 samples with V_p decreasing but also the others (such as granites). All the 54 samples except for marbles contains horblende or biotite, and the phase-transition from horblende to pyroxene was also found.

Dehydration has positive effect on partial melting. For instance, in granites of our experiment. The temperatures and pressures have been over the liquidus, but there is not any melting glasses in the products based on our observation. V_p decrease phenomenon was not be found in the granites yet. Water, ether existing in the water-bearing minerals or we add into, is very important for partial melting in rocks. The samples with V_p decrease phenomenon all contain water-bearing minerals (horblende or biotite), dehydration occurred before and been the prerequisite factor of partial melting. Melting decrease V_p further. Such as, the low velocity bodies tested by CT in Yellow Stone Park and East Africa Rift Belt were interpreted as partial melting of deep crust [15].

The start conditions of dehydration are the pressure and temperature corresponding to V_{Pmax}, about 700-800 ℃ and 0.6-0.7GPa for the rocks containing hornblende. The conditions corresponding to V_{Pmin} may mean considerable melting has been in the rocks.

The discussion above mentioned implies that dehydration, phase-transition and partial melting are really the reason for the V_p decrease in rocks in our experiments. Combined other studies that the conductivities will increase appearly during dehydration and phase-transition [5], both the lower velocity and high conductivity of rocks could be caused by the process talked above.

A comparison of V_p decrease in rocks and in the crust must be done when we talk about the origin of the crustal low velocity layer in Qinling and North China. Amphibolite ficies rocks (such as amphibolite, meta-basic vocanic rocks) are dominated, and the water-bearing minerals (such horblende and biotite) are aboundant, in the middle-lower crust of the Qinling Orogenic Belt, Inner Mongolia and Wutaishan [6, 16]. The study areas has been confirmed to be high heat flow regions. Shown by the geotherm of the regions [7], the start temperature (700 ℃) and pressure (0.6GPa) of dehydration could be reach in the middle-lower crust. Two possibilities may be exsit, one is that dehydration, phase-transition and partial melting happened in the middle crust and caused the CLVL. The other one is as the results studied by Etheridge [3], dehydration, phase-transition and partial melting happened in the lower crust. The supercritical water and melting material upwelled to the middle crust and traped by a cover, hence caused the CLVL. Therefore, the dehydration (dominented by Horblende), phase-transition and partial melting would be one of the most possible candidates to answer the origin of the CLVL of the study areas.

CONCLUSIONS

Mineral constitutions and textures of the 54 experimental products indicate that the dehydration, phase transition and partial melt of water-bearing minerals, such as hornblende and biotite, are responsible for the V_p decrease. The results suggest that dehydration melting of hydrous minerals might be an important mechanism for CLVL observed in the continental crust of the study area and probably other parts of the world.

In order to find the oringin of both low velocity and high conductivity of CLVL, a progressive work needs to do in the future. We plan to measure P-wave velocities and electrical parameters simultaneously in the same sample, and do some parallel experiments to observe the change in products at various P-T conditions.

Acknowledgements

We thank assistant Professer Jianping Ouyang, Hongfei Zhang, Jifeng Xu and Dr. Wenge Zhou for their field assistance. Some helpful discussion with assistant Professer Qingsheng Liu. Special thanks to Professer Jinfu Deng and Professer F. Wenzel for their warmly help. This work is funded by the projects provided by the National Natural Science Foundation of China (NSFC) and the Ministry of Geology and Mineral Resources of China (MGMRC).

REFERENCES

1. N.I. Christensen. Compressional wave velocities in rocks at high temperatures, critical thermal gradients, and crustal low-velocity zones, *J. Geophys. Res.*, 84, B12, 6849-6857 (1979).
2. Z.Z. Cui, Z.X. Yin, E.Y. Gao, et al.. Velocity structure and deep-seated structure of the Qinghai-Xizang (Tibet) Plateau, Geological Publishing House, Beijing (1993) (in Chinese, with English abstract).

3. M.A. Etheridge, V.J. Wall, S.F. Cox, et al. The role of the fluid phase during regional metamorphism and deformation, *J. Metamorph. Geol.*, 1, 205-226 (1983).

4. D.M. Fountain. The Ivrea-Verbano and Strona-Ceneri zones, northern Italy: A cross section of the continental crust – new evidence from seismic velocities, *Tectonophysics*, 33, 145-165 (1976).

5. P. Gao, R.X. Liu, B.L.Ma, B. Li and R.C. Mu. The physical and mechanical properties of chlorite-schist and plagioclase amphibolite at high P T conditions and its application, *Seismology and Geology*, 16 (1), 83-88 (1994) (in Chinese, with English abstract)

6. S. Gao, B.R. Zhang, T.C. Luo, et al. Chemical composition of the continental crust in the Qinling Orogenic Belt and its adjacent North China and Yangtze cratons, *Geochim.Cosmochim.Acta.*, 56, 3933 – 3950 (1992).

7. S. Gao and B. Zhang. Radioactivity of rocks in the Qinling Orogenic Belt and adjacent areas and the current thermal structure and state of the lithosphere, *Chinese Journal of Geochemistry*, 3, 241-251 (1993) (in Chinese, with English abstract),

8. R.D. Hyndman, L.L. Vanyan, G. Marquis, et al., The origin of electrically conductive lower continental crust: saline water or graphite? *Phys. Earth Planet. Int.*, 81, 325-344 (1993)

9. B.H. Jiang, D.W. Yuan, Y.H. Wu. Relation of low-velocity crustal layers with earthquake activities, *Geological Science and Technology Information*, 11(4), 9-15 (1992) (in Chinese, with English abstract)

10. H. Kern. The effect of high temperature and high confining pressure on compressional wave velocities in quartz-bearing and quartz-free igneous and metamorphic rocks, *Tectonophysics*, 44, 185-203 (1978).

11. H. Kern and V. Schenk. A model of velocity structure beneath Calabria, southern Italy, based on laboratory data, *Earth Planet. Sci. Lett.*, 87, 325 – 337 (1988).

12. S. Mueller and M. Landisman. Seismic studies of the Earth's crust in continents, I, Evidence for a low velocity zone in the upper part of the lithosphere, *Geophys. J.*, 10, 525-538 (1966)

13. D.H. Shurbet and S.E. Cebull. Crustal low-velocity layer and regional extension in Basin and Range Province, *Geological Society of America Bulletin*, 82(11), 3241-3243 (1971)

14. W.C. Sun, S.L. Li, L.L. Luo and H.F. Yue. A preliminary study on low velocity layer in the crust in North China, *Seismology and Geology*, 9(1), 17-26 (1987) (in Chinese, with English abstract).

15. X.Y. Wang, R. Gao, J.F. Deng, et al.. Review on The International Lithosphere Program (ICP). In: *Studies on the crust and upper mantle*, R.J. Xiang, C.Z. Shi and Z.X. Feng (Ed.). pp. 14-24. Seismology Press, Beijing (1991) (in Chinese, with English abstract).

16. Z.X. Wu, J.F. Deng, H.L. Zhao, et al.. Crust-mantle petrological structure and evolution of the North China continent, *Acta Petrologica et Mineralogica*, 13(2), 106-115 (1994) (in chinese, with English abstract)

17. H.S. Xie, Y.M. Zhang, H.G. Xu, et al. A new method of measurement for elastic wave velocities in minerals and rocks at high temperature and high pressure and its significance, *Science in China (Series B)*, 36(10), 1276-1280 (1993)

18. J.A. Xu, Y.M. Zhang, W. Hou, H.G. Xu, J. Guo, Z.M. Wang, H.R. Zhao, R.J. Wang, E. Huang and H.S. Xie. Measurements of ultrosonic wave velocities at high temperature and high pressure for window glass, pyrophyllite, and kimberlite up to 1400 $_{℃}$ and 5.5 GPa, *High Temperatures-high Pressures*, 26, 375-384 (1994)

19. X.C. Yuan. Deep structure and tectonic evolution of Qinling orogenic zone. In: *A selection of papers presented at the conference on the Qinling Orogenic Belt*, L.J. Ye, X.L.Qian and G.W.Zhang (Ed.). pp. 174-184, NW Univ. Press, Xi'an (1991) (in Chinese, with English abstract).

20. Z.D. Zhao. *Geochemistry of the Yichuan (Henan) – Yichang (Hubei) Geoscience Transect, East Qinling, China* [unpublished Ph.D. Thesis]. China University of Geosciences, Wuhan (1995) (in Chinese, with English abstract).

Proc. 30th Int'l. Geol. Congr., Vol. 4, pp. 31-43
Hong Dawei (Ed)
© VSP 1997

Elastic Properties of Crust and Upper Mantle — A New View

FRIEDEMANN WENZEL, KARL FUCHS, UWE ENDERLE AND MARC TITTGEMEYER
Geophysical Institute, Karlsruhe University, Hertzstr. 16, 76187 Karlsruhe, Germany

Abstract

A seismological characterization of crust and upper mantle can refer to large scale averages of seismic velocities or to fluctuations of elastic parameters. Large is understood here relative to the wavelengths used to probe the earth. For the lower crust it is common to study both properties by combining reflection and refraction seismology. For the crust/mantle transition the fluctuations are reflected in the coda properties of the critical and postcritical wide-angle Moho reflection. For the uppermost mantle the long-range (teleseismic) P_n- and S_n-waves carry crucial information.

Keywords: lower crust, upper mantle, seismology

INTRODUCTION

Seismologists have two ways to characterize the part of the lithosphere that consists of lower crust, crust/mantle transition (Moho) and uppermost mantle. They either specify the elastic velocities, or its fluctuations.

In the first case the earth is described by a velocity-depth function, which typically shows P-wave velocities between 6.4 and 7.2 km/s in the lower crust, a jump to velocities between 8.0 and 8.3 km/sec at Moho level, and a small gradient of increasing velocities in the upper mantle. This velocity-depth function may change laterally in a given terrain and will show modification in values for different tectonic regimes. The velocity models are derived from refraction seismic data. The elastic waves generated with controlled seismic sources are received at the surface along seismic lines, which extend from the source to distances of 150 to several hundreds of km. The waves that propagate through the crust and upper mantle along different travelpaths appear as seismic phases in the seismogram sections. The traveltimes of those phases can be inverted for the velocity-depth structure.

Inversion of traveltimes from correlated phases results in a model where velocities between interfaces are averaged over many wavelengths. In fact the physical notion of a phase relies on ray theory which is only applicable for media with large blocks of velocities which do not change strongly on the scale of a wavelength. The individual blocks (or layers) can be separated by sharp interfaces [2].

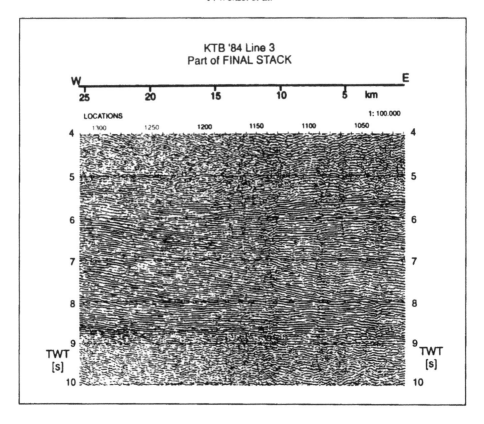

Figure 1. Lower crustal seismic lamellae as observed on a seismic reflection profile recorded in the Black Forest, SW Germany [13]. Crustal reflectivity increases abruptly at about 5.5 s two-way time (TWT) and terminates at 9 s at the Moho. The high amplitudes of the reflections require the existence of significant velocity contrasts (±5% or larger) in the lower crust. Lower crustal seismic lamellae have been observed throughout the world with varying amplitudes, lateral continuity, and dips depending on the tectonic setting.

Fluctuations of the medium appear as wave scattering and cause backscattering or coda effects in transmitted waves [1]. Scattering analysis is widely done for crustal earthquakes and array records of teleseismic events, but rarely for controlled source data others than near-vertical reflection data. After a review on the properties of the lower crust the study of coda properties of the wide-angle Moho reflection (P_MP) and long-range P_n/S_n-waves is used to demonstrate how scales of fluctuations can be seen and determined.

CHARACTERISTICS OF THE LOWER CRUST

Important information on the continental lithosphere are provided by near-vertical reflections backscattered from the lithosphere along reflection seismic profiles. These data carry little information on the crustal velocities but allow to generate seismic section - either stacked sections or migrated sections - which provide an image of the velocity and/or density fluctuations. If the earth were as homogeneous as most models inverted from refraction seismic experiments suggest reflection seismology had never developed for lithospheric applications. However, the earth's crust does contains fluctuations within the crust, which can be imaged by reflection seismology. In the upper crust these reflections can be correlated with features of surface geology and represent deep sedimentary layers, volcanic intrusions, thrust and normal faults. In general faults become listric at midcrustal levels and never cut through the lower crust. The latter is very often characterized by numerous sub-horizontal or shallow dipping reflections, which seem to be randomly distributed in the lower crust. They terminate abruptly at a depth, which usually coincides with the Moho as found from wide-angle refraction data. This widespread observation gave rise to the notion of laminated or layered lower crust. A variety of explanations were initially discussed as physical cause for these reflection pattern. The most widely accepted is that the layered lower crust is the result of intensive stretching under thermodynamic conditions where rocks deform in plastic flow. This flow affects material of different impedance and generates anisotropic orientation of minerals by localized deformation. This effect would produce the observed predominantly horizontal reflection pattern. Fig. 1 shows a reflection section from the central Black Forest [13]. Its reflectivity is concentrated within the lower crust; the upper mantle appears as transparent; only few reflections are discernible in the upper crust.

Velocity depth distribution and reflectivity pattern carry different information. The first set of data reflects the state of chemical differentiation of the earth and the metamorphic conditions at depth. If models derived from refraction data are grouped according to the tectonic setting one finds characteristic differences between old shields, rifted areas, crust thickened by ongoing orogeny, volcanic arcs a.s.o. [19].

Reflectivity of the lower crust is usually attributed to extension and/or magmatic intrusions. Extension can be a result of late to post-orogenic collapse or active or passive rifting of the lithosphere. In both scenarios lower crustal temperatures are high enough to promote plastic flow which tends to align minerals and material of different composition horizontally. The dynamic processes that affect the lower crust generate a particular scale (vertical and horizontal) that can be observed with reflection seismic experiments. They generally invoke frequencies between 10 and 50 Hz and near-vertical wave propagation.

The lower crustal reflectivity and wide-angle observations are related in two ways. The velocity (or impedance) fluctuations within the lower crust are also visible in high-resolution refraction data. They appear as coda wavetrain that follows the midcrustal reflection from the Conrad discontinuity. This has been observed from and modelled for the same data set shown in Fig. 2 [25]. It served as proof that impedance fluctuations are actually causing the reflections and allowed to quantify the size of fluctuations.

The second more surprising feature is that the bottom of the reflectivity pattern usually coincides with the Moho as defined from refraction data. This has been shown for many coincident reflection/refraction lines [20]. Both data sets become comparable if a depth

F. Wenzel et al.

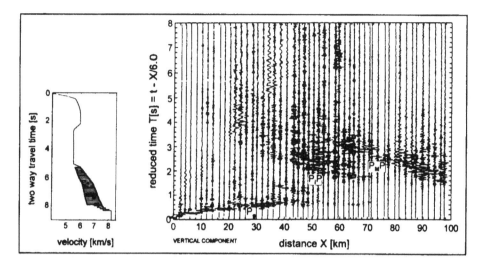

Figure 2. Record section of the refraction profile in the Black Forest. The panel shows the vertical component. Reduction velocity is 6.0 km s^{-1}. Upper crustal phase (P$_g$) and mantle reflections (P$_M$P) are recognized. Data are scales to maximum amplitude. The left panel shows the preferred 1D model with strong fluctuation in the lower crust causing the P$_i$P coda.

conversion of reflection or a time conversion of refraction data is performed. The generally accepted explanation is that the lower crust deforms in a plastic way whereas the uppermost mantle due to its different lithology may be brittle. However, this does not explain the generation of the heterogeneities and their disappearance at Moho. Another possible explanation is that lower crust and upper mantle deform in a ductile way but different lithologies and/or different strain rates generate different scales of structures. Those developed in the lower crust are detectable with the near-vertical reflection seismic geometry and frequency window; those in the upper mantle are not. In studies of the crust and upper mantle the most widely used method is controlled source seismology (reflection or refraction studies). Records of the transient wave field of teleseismic earthquakes are only recently inspected to detect small scale fluctuations in velocity [23] and widen the freqency spectrum at the lower end from 10 Hz to about to 2 Hz.

From reflection records the notion of reflective lower crust and transparent upper mantle developed. It must be stressed, however, that the mantle is only transparent for the frequency range and the geometry of this particular method, namely for near-vertical incidence. There is no reason to believe that the mantle is homogeneous on a scale of tens of kilometres, that it has not been affected by tectonic processes and that those processes have not left their fingerprints as impedance fluctuations on a particular scale. In fact inspection of wide-angle data and their coda properties reveal that a significant change in scale of fluctuations occurs especially between lower crust and upper mantle [4]. The particular scale at Moho level is imaged in the coda properties of the wide-angle Moho reflection. The scale of mantle fluctuations is contained in long-range refraction data which allow to observe the so-called teleseismic P$_n$ and S$_n$ phases.

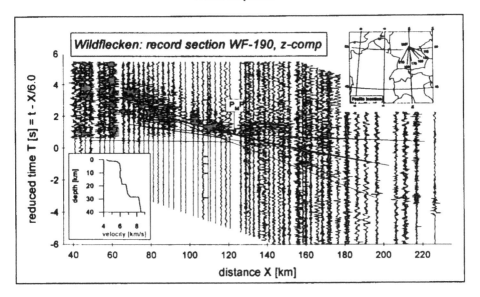

Figure 3. Crustal record section of the Wildflecken seismic refraction experiment in southern Germany [31]. The data shown here are plotted trace-normalized with a reduction velocity of 6 km/s. Note the predominance of the wide-angle P_MP reflection from the crust-mantle boundary and its short but distinct coda. Additionally, travel-time branches computed from a 1D-model (shown in the inset) are drawn into the record section.

SCALES OF STRUCTURE AT MOHO LEVEL

In most reflection sections the Moho is marked as the end of the reflective lower crust. Although exceptions exist in general the Moho does not show up as a particular reflective zone. From refraction data we are familiar with the well-known velocity discontinuity which usually generates a strong wide-angle Moho reflection (P_MP). Combining both observations could lead to the conclusion that Moho is a strong (usually 1st order) velocity step, which is associated with the bottom of the velocity fluctuations of the lower crust. If this were the case one would observe (1) a strong wide-angle reflection, (2) reverberations after the mid-crustal Conrad reflection (P_iP), and (3) no strong coda around the critical P_MP distance.

The refraction record shown in Fig. 3, however, shows a conspicuous coda appended to the reflection curve. It represents a profile from SW-Germany (Wildflecken: [31]). Data are reduced in time by 6.0 km/s and traces are plotted normalised. Although it consists of a few cycles the P_MP coda is quite prominent. The record section of Fig. 3 is a good example of a very common feature. Such strong P_MP coda is visible on a variety of similar refraction data sets (Black Forest: [9]; Saxonian Granulite Range: [5]; NW-France: [11]; Massif Central: [32]; LISPB: [6]; Fennolora: [26]; KRISP: [22]; Jordan: [15]).

While the travel time of the P_MP phase is used to determine the Moho depth, the coda of the supercritical P_MP provides a very important constraint for the properties of the crust-

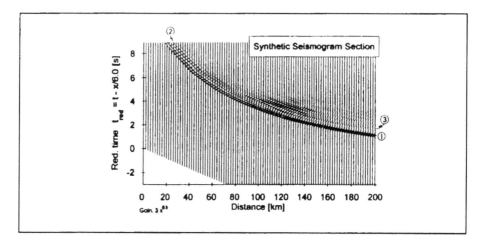

Figure 4. Synthetic seismogram section (right) calculated with the reflectivity method [7] for a simple 1D-model (left) to demonstrate the effect of the 'supercritical resonator' (SCR). Traces are plotted with true amplitudes in a reduced time scale (V_{red} = 6.0 km/s) and a distance-dependent amplification of $x^{0.5}$ to compensate for geometrical spreading of energy.

mantle transition. Considering the variability of observed seismic record sections and the extreme complexity of the crust in general, this P_MP with its characteristic coda is the most consistent and pronounced observation in western Europe, but also on other continents. Therefore, any model for the crust-mantle transition must reproduce the P_MP with its coda.

Enderle et al. [4] argued that the observed P_MP-coda must be regarded as a wave package of supercritical multiple reflections or peg-legs with about the same amplitudes as the primary P_MP reflection, slightly shifted in space and displaced in time by about the dominant period of the primary P_MP reflection. To model such coda properties, requires the presence of a particular resonator in the depth range of the crust-mantle transition zone. Enderle et al. [4] call it a 'supercritical resonator' (SCR).

It consists of a series of high-velocity (HVL) and low-velocity layers (LVL). The essential element of the SCR is a combination of appropriately tuned HVL and LVL at the depth of the classical Moho. Many numerical experiments with the aim of finding a one-dimensional model for the P_MP coda with the reflectivity method [7] showed that only a combination of HVL-velocities as high as upper mantle material (e.g., about 8.0 km/s) and of LVL-velocities as low as 6.0 km/s (felsic crustal material, possibly in some cases even partial melt) are required for such a tuned 'resonator'. One successful model of the SCR with the synthetic P_MP-coda is shown in Fig. 4.

The high velocity layers (HVL) of the SCR provide strong supercritical reflections from heterogeneities within the transition zone with much less decay of amplitude than at precritical incidence. The interbedded low velocity layers (LVL) provide the necessary delay. The lower the velocity in the LVL the steeper is the ray path and the smaller is the offset produced for the multiple reflections.

Thus the classical definition of the crust/mantle transition as substantial velocity increase is supplemented by the description as a change in scales of structure: Interlayering of very high (mantle) and very low (crustal) velocity material on the particular vertical scale discussed above. A statement on the lateral extent of these fluctuations is not yet possible. We assume, however, that the lateral correlation length is within the range of 5 km (Fresnel zone) in order to guarantee the evolution of specular reflections and high coda energy. The principal velocity contrast of the Moho appears as high amplitude P_MP reflection. The fluctuations determine the supercritical P_MP coda.

FLUCTUATIONS IN THE UPPER MANTLE

Controlled source seismology (reflection and refraction data) provide the impression that the uppermost mantle is rather homogeneous. For near-vertical reflections the mantle appears transparent as compared to the reflective lower crust. Mantle reflections are sometimes observed, but as discrete reflections rather than spatially distributed reflectivity. Refraction data allow the observation of a P_n phase which serves to define a sub-Moho velocity or velocity gradient. Long-range observations allow to define discrete velocity contrasts in the mantle, but again the models display a mantle which is mostly homogeneous on the scale of many wavelengths.

Description of the uppermost mantle by its velocity fluctuations rather than by its average velocity are restricted to interpretation of studies of wave fluctuation beneath seismological arrays [30], to modelling waveforms for propagation in oceanic lithosphere [10] and more recently to the understanding of a high-frequency phase observed in Russian long-range data with nuclear explosions as source. These long-range refraction data from the Russian seismological Peaceful Nuclear Explosion (PNE) Program indicate the existence of the so called teleseismic P_n phase [24], which travels within the upper mantle lithosphere to observational distances of several thousands of kilometers. This phase can clearly be identified on the shots of the PNE profile QUARTZ (Fig. 5). The notion teleseismic P_n has been used previously by Molnar and Oliver [18], who investigated efficient long-range propagation of mostly S_n waves generated by earthquakes. Those waves could be observed to travel with upper mantle velocities for several thousand of kilometers and are reported as high frequency signals both for continental and oceanic paths. Unfortunately the sparseness of the net of earthquake stations did not allow to plot seismic sections and thus to visualize the evolution of the teleseismic P_n with distance.

As a likely explanation for the efficient propagation of the P_n several authors consider the existence of a sub-Moho waveguide. For this waveguide several theories have been proposed: Menke and Richards [16], Stephans and Isacks [27], Mantoviani *et al.* [14], Sutton and Walker [28], Fuchs and Schulz [8], Menke and Richards [17]. Ryberg *et al.* [24] and Tittgemeyer *et al.* [29] showed that a waveguide, which is generated by random

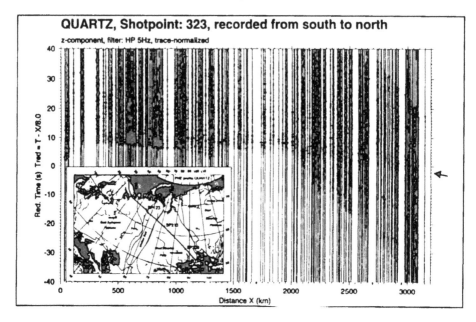

Figure 5. High-pass filtered (corner frequency f = 5 Hz) vertical-component time-distance record section (reduction velocity 8.0 km/s) on profile QUARTZ (for the location see inset) for shotpoint 323 recorded to the Northwest. It is trace normalized individually to maximum trace amplitude. The marked strong wave band travelling with a group velocity of 8.1 km/s is the high-frequency teleseismic P_n.

successions of high and low velocity layers can quantitatively explain the main features of the teleseismic P_n as observed on profile QUARTZ such as coda length, enhancement of high frequencies, and a group velocity of 8.0 to 8.1 km/s.

The data analysed for the high-frequency teleseismic P_n phase belong to the system of long-range deep seismic sounding profiles carried out by Russian scientists from 1971 to 1990. Peaceful Nuclear Explosions (PNE) and numerous chemical explosions were used as powerful sources. In particular, the seismic data from the QUARTZ profile, located in northern Eurasia, were used to investigate the high-frequency teleseismic P_n. About 400 short period (1-2 Hz) three component, analogue recording systems were deployed to record the ground motion resulting in maximum observation distances of about 3200 km. The average station spacing of about 10 km along the profiles provides a unique data density for studies of the teleseismic P_n.

The recorded wavefield of the 3 main shots on the QUARTZ profile shows distinct phases that are associated with the large scale velocity structure of the crust and upper mantle. Recent studies on the velocity structure below northern Eurasia show remarkable lateral inhomogeneities in the crust and upper mantle [3]. In addition to these distinct phases, the high-frequency part of the wavefield differs strongly from the corresponding low-frequency constituent. Ryberg et al. [24] show, that it is clearly dominated by the teleseismic P_n, defined as a phase which travels with a group velocity of 8.1 km/s and has no sharp first onset. At distances greater than 1300 km it arrives after the wave diving through the upper mantle. This phase contains only high-frequency energy in the band between 5-12 Hz and is characterised by a long incoherent coda.

Our favourite upper mantle model is shown in Fig. 6. It includes a standard crustal structure with homogeneous upper and lower crust with P wave velocities of 6.0 km/s for the upper and 6.5 km/s for the lower crust between 16 and 35 km depth. Moho is marked by a first order discontinuity at 35 km depth. The upper 75 km of the mantle are characterised by a slightly positive gradient zone. For this zone we assume a constant Q_P of 1400 and a constant Q_s of 600.

The observations on long-range seismic profiles call for a class of upper mantle models which are characterised by random velocity fluctuations. If they are in the range of $\pm 4\%$ a new feature in the wavefield, the teleseismic P_n, arises. Because the actual size of fluctuations will probably change with the dimensionality of the models we do not want to discuss petrological implications in a quantitative sense. It is conceivable that the fluctuations represent preferred orientation of olivine with a random component.

If the type of layering we see in PNE data is a more general phenomenon as speculated by Enderle *et al.* [4] it could be interpreted as remnant of a disturbed laminar flow

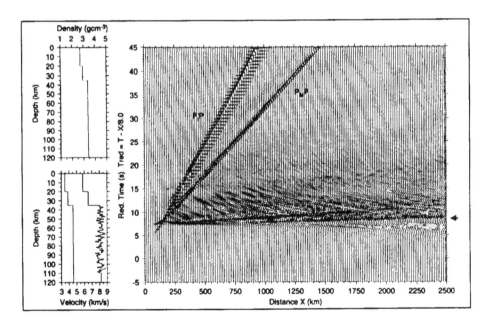

Figure 6. P-wave theoretical record section calculated with the reflectivity method. The reduction velocity is 8.0 km/s. The velocity-depth functions and the density-depth function is plotted to the left. The P-wave velocity fluctuations which are superimposed on the positive upper mantle gradient are $\pm 4\%$. The correlation length is 2 km. Note that a teleseismic P_n is generated which is characterized by a group velocity of 8.0 km/s and a extensive coda.

pattern underneath the Moho that must be different from material flow in the lower crust. The latter is generally assumed to result in late- to post-orogenic settings where crustal stacking in a high temperature environment is re-equilibrated by intensive lower crustal extension, which is ultimately responsible for its observed reflectivity pattern. Upper mantle and lower crust are different in terms of composition, compositional variety and rheology. Lower crust is assumed to be predominantly mafic but to contain both mafic and felsic rocks which allow for high impedance contrasts. The mantle could have a mineralogically more homogeneous composition but with highly anisotropic material in particular olivine. Fluctuations caused by anisotropy are pure velocity variations with unchanged densities. This would result in fairly small near vertical impedance contrasts. Also mantle material at a given geotherm can yield higher differential stresses than lower crustal rocks. These differences may result in different scales of vertical layering and horizontal extend of inhomogeneities.

CONCLUSIONS

The data presented so far indicate that the crust/mantle system shows conspicuous fluctuations in elastic parameters: Reflection data show a layered or reflective lower crust with velocity fluctuations as high as 10%, with variable vertical and lateral scales, the latter ones typical being in the km range. At the crust/mantle boundary (Moho) the scale of inhomogeneities changes. The P_MP coda requires stronger fluctuations (20%) which may represent a mixture of crustal and mantle material. We present arguments that a larger horizontal correlation length can be expected though this has not yet been quantified by modelling. As several data sets from various tectonic regimes show a strong P_MP coda one could speculate that this is as global as the Moho if defined as substantial inversion in velocity at about 40 km depth.

Superlong controlled source data from Russia give evidence for random fluctuations in the upper mantle. 1D-modelling provides a vertical scale of the inhomogeneities inn the km-range. The horizontal correlation length is unknown. We believe that it must be in the 10 km range in order to trap the elastic energy and propagate it out to 3000 km offset from the source. Observations of S_n waves recorded at permanent seismic stations and data with oceanic travel path indicate that this upper mantle fluctuations may also represent a wide-spread feature. Fig. 7 shows a cartoon which summarizes the described properties. It is schematic because fluctuations are shown as regular rather than as random features, which is their likely nature. We choose this representation in order to indicate the different scales more clearly.

Once the scales and their variations are established their geological significance becomes the key problem. At this point only speculations and options can be discussed. The layered lower crust may represent the evidence for a decoupling of motion between upper crust and mantle during tectonic processes acting at high temperatures. The Moho with its conspicuous crust/mantle mix could have acted as detachment during these processes. This statement is identical with saying that the largest change in strain rate occurred across the Moho. Thus the commonly used model of constrained strain rate

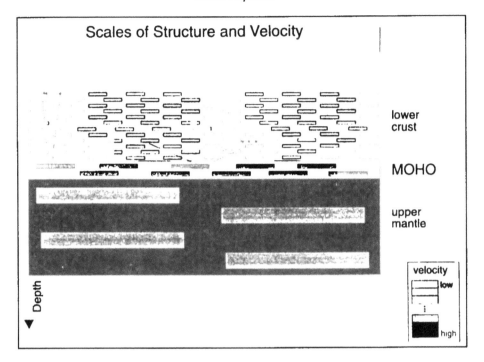

Figure 7. Conceptional diagram indicating the changing scale of fluctuations from crust to mantle. Small scale heterogeneities in the lower crust are responsible for the reflectivity of the lower crust as seen in near-vertical seismic experiments. The difference of refraction and reflection Moho produces the various reflective patters as observed in near-vertical and wide-angle seismic experiments. The zone below reflection Moho with its mildy fluctuating large-scale structures serves as a wave guide for the teleseismic P_n phase.

beneath the mid-crust [12] must be revised. A constant stress assumption is more appropriate [21]. Inhomogeneities in the uppermost mantle with large lateral extent may be remnants of mantle flow processes.

REFERENCES

1. K. Aki and B. Chouet. Origin of coda waves: Source, attenuation, and scattering effects. *J. Geophys. Res.* 80, 23, 3322-3342 (1975).
2. K. Aki and P.G. Richards (Eds). *Quantitative seismology*, vol. 2, W.H. Freeman, San Francisco (1980).
3. V.A. Egorkin and A. V. Mikhaltsev. The results of Seismic Investigations Along Geotraverses. In: *Super-Deep Continental Drilling and Deep Geophysical Sounding.* K. Fuchs, Y. A. Kozlovsky, A. I. Krivtsov, M. D. Zoback (Eds). pp. 111-119, Springer Verlag Berlin Heidelberg New York (1990).
4. U. Enderle, M. Tittgemeyer, M. Itzin, C. Prodehl and K. Fuchs. Scales of structure in the lithosphere - images of processes. In: *Stress and Stress Release in the Lithosphere.* K. Fuchs, R. Altherr, B. Müller, and C. Prodehl (Eds). subm. Tectonophysics (1996).
5. U. Enderle, K. Schuster, C. Prodehl, A. Schulze and J. Briebach. The refraction seismic experiment GRANU95 in the Saxothuringian belt, SE-German, in preparation (1996).
6. S. Faber and D. Bamford. Lithospheric structural contrasts across the Caledonides of northern Britain. In: *Structure and Compositional Variations of the Lithosphere and the Asthenosphere.* K. Fuchs and M.H.P. Bott (Eds). pp. 17-30, Tectonophysics, 56 (1979).
7. K. Fuchs and G. Müller. Computation of synthetic seismograms with the reflectivity method and comparison with observations. *Geophys. J.R. Astron. Soc.* 23, 417-433 (1971).
8. K. Fuchs and K. Schulz. Tunnelling of low-frequency waves through the subcrustal lithosphere. *J. Geophys.* 42, 175-190 (1976).
9. D. Gajewski and C. Prodehl. Seismic refraction investigation of the Black Forest. *Tectonophysics* 142, 27-48 (1987).
10. J.F. Gettrust and L.N. Frazer. A computer model study of the propagation of long-range P_n phase. *Geophys. Res. Lett.* 8, 749-752 (1981).
11. A. Hirn, L. Steinmetz, R. Kind and K. Fuchs. Long range profiles in western Europe - II. Fine structure of the lower lithosphere in France (southern Bretagne). *Z. Geophys.* 39, 363-384 (1973).
12. N.J. Kusznir. The distribution of stress with depth in the lithosphere: thermo-rheological and geodynamic constraints. *Philosophical Transactions of the Royal Society of London* Series A, 337(1645), 95-110 (1991).
13. E. Lüschen, F. Wenzel, K.-J. Sandmeier, D. Menges, Th. Rühl, M. Stiller, W. Janoth, F. Keller, W. Söllner, R. Thomas, A. Krohe, R. Stenger, K. Fuchs, H. Wilhelm and G. Eisbacher. Near-vertical and wide-angle seismic surveys in the Black Forest, SW Germany. *J. Geophys.* 62, 1-30 (1987).
14. E. Mantovani, F. Schwab, H. Liao and L. Knopoff. Teleseismic S_n: a guided wave in the mantle. *Geophys. J.R. Astron. Soc.* 51, 709-726 (1977).
15. J. Mechie and Z.H. El-Isa. Upper lithospheric deformations in the Jordan-Dead Sea transform regime. In: *The Gulf of Suez and Red Sea Rifting.* X. LePichon and J.R. Cochran (Eds). pp. 103-121, Tectonophysics 153 (1988).
16. W.H. Menke, and P. G. Richards. Crust-mantle whispering gallery phase: A deterministic model of teleseismic P_n propagation. *J. Geophys. Res.* 85, 5416-5422 (1980).
17. W.H. Menke, and P. G. Richards. The horizontal propagation of P waves through scattering media: Analog model studies relevant to long-range P_n propagation. *Bull. Seismol. Soc. Am.* 73, 125-142 (1983).
18. P. Molnar, and J. Oliver. Lateral variations of attenuation in the upper mantle and discontinuities in the lithosphere. *J. Geophys. Res.* 74, 2648-2682 (1969).
19. W.D. Mooney and R. Meissner. Multi-generic origin of crustal reflectivity: A review of seismic reflection profiling of the continental lower crust and Moho. In: *Continental Lower Crust.* D.M. Fountain, R. Arculus and K.W. Kay (Eds). pp. 179-199. Elsevier, Amsterdam. (1992).
20. W.D. Mooney and T.M. Brocher. Coincident seismic reflection/refraction studies of the continental lithosphere: a global review. *Rev. Geophys.* 25, 723-742 (1987).
21. B. Müller, V. Wehrle, H. Zeyen and K. Fuchs. Short-scale variations of tectonic regimes in the Western European stress province north of the Alps and Pyrenees. In: *Stress and Stress Release in the Lithosphere.* K. Fuchs, R. Altherr, B. Müller, and C. Prodehl (Eds). subm. Tectonophysics (1996).
22. C. Prodehl, G.R. Keller and M.A. Khan. Crustal and upper mantle structure of the Kenya rift. *Tectonophysics* 236, 483p. (1994).
23. J.R.R. Ritter, M. Kempter, G. Stoll and K. Fuchs. Scattering of teleseismic waves in the lower crust - observations in the Massif Central, France. subm. *Phys. Earth Planet. Inter.* (1996).

24. T. Ryberg, F. Fuchs, A. V. Egorkin and L. Solodilov. Observations of high-frequency teleseismic P_n waves on the long-rang QUARTZ profile across Northern Eurasia. *J. Geophys. Res.* 100, 18151-18163 (1995).

25. K.-J. Sandmeier, and F. Wenzel. Synthetic seismograms for a complex model. *Geophys. Res. Lett.* 13, 22-25 (1986).

26. R. Stangl. Die Struktur der Lithosphäre in Schweden, abgeleitet aus einer gemeinsamen Interpretation der P- und S-Welle Registrierungen auf dem FENNOLORA-Profil. Ph.D. thesis, Karlsruhe University, 187p. (1990).

27. C. Stephans, and B. L. Isacks. Toward an understanding of S_n; Normal modes of Love waves in an oceanic structure. *Bull. Seismol. Soc. Am.* 67, 69-78 (1977).

28. G.H. Sutton, and D. A. Walker. Oceanic mantle phases recorded on seismographs in the northwestern Pacific at distances between 7° and 40°. *Bull. Seismol. Soc. Am.* 62, 631-655 (1972).

29. M. Tittgemeyer, F. Wenzel, K. Fuchs and T. Ryberg. Wave propagation in a multiple scattering upper mantle - observations and modelling. *J. Geophys. Int.* in print (1996).

30. R.-S. Wu, R and S.M. Flatté. Transition fluctuations across an array and heterogeneities in the crust and upper mantle. *PAGEOPH* 132, 1/2 (1990).

31. St. Zeis, D. Gajewski and C. Prodehl. Crustal structure of Southern Germany from seismic-refraction data. *Tectonophysics* 176, 59-86 (1990).

32. H. Zeyen, O. Novak, M. Landes, C. Prodehl, L. Driad and A. Hirn. Refraction-seismic investigations of the northern Massif Central (France). In: *Stress and Stress Release in the Lithosphere*. K. Fuchs, R. Altherr, B. Müller, and C. Prodehl (Eds). subm. Tectonophysics (1996).

Proc. 30th Int'l. Geol. Congr., Vol. 4, pp. 45-52
Hong Dawei (Ed)
© VSP 1997

A Combined Genetic and Nonlinear Analytic Inversion Method for Anisotropic Media

ZHOU HUI

Institute of Geology and Geophysics, Ocean University of Qingdao, 266003, P. R. China.

HE QIAODENG

Department of Geophysics, Changchun University of Earth Sciences, 130026, P. R. China.

Abstract

The inversion of anisotropic parameters generally is a nonlinear problem of optimization. Nonlinear inversion methods can be classified two groups: random walk methods, such as Monte Carlo method, simulated annealing and genetic algorithm (GA); and the other group of nonlinear analytic inversion (NAI) method based on nonlinear least-squares theory. GA is able to find global optimal solution, but it is a time consume technique. NAI can give a local optimal solution which relates to the starting model, however, the convergence rate of this method is high. The purpose of the combined genetic and nonlinear analytic inversion method is to utilize the advantages and to overcome the shortcomings of the two inversion methods. So the combined inversion (CI) method would not only be able to converge quickly, but to find global optimal solution. It is important how to properly combine GA and NAI. We propose the combining scheme would be that after several consecutive GA iterations one time of NAI is done. Theoretical examples demonstrate that CI has the characteristics mensioned above.

Keywords: genetic algorithm, nonlinear analytic inversion method, combined nonlinear inversion method, anisotropic media

INTRODUCTION

Genetic algorithm, one of random exploration methods, is a global search method based on probabilistic mechanisms for solving complex optimization problems with multiple minima [9]. GA does not require that the shape of the posterior probability density in the model space be known and extensively samples the model space. GA is established with an analogy to the mechanics of natural selection and genetics and based on the principle of survival of the fittest.

GA is proposed by Holland in 1975 [4]. Berg [1] firstly applied GA to geophysical optimization problems. Stoffa et al. [11] systematically studied the effects of the size of population, probability of crossover, selection and mutation on the convergency and convergence rate for multiple parametric optimization problems. Sen et al. [10] discussed the function of temperature introduced in probability of selection and suggested some annealing schemes. Zhou et al. [16] did some research work on the relationship between objective function and convergence rate and accuracy of solution. Zhou [17] gave a new coding method for large multiple parametric optimization problems.

The nonlinear inversion techniques based on least-squares optimization theory is one of two groups. This descent type algorithms give the correct solution when starting model is inside the valley of the global minimum of objective function, regardless of the choice of starting point. Therefore, the descent type algorithms are in widespread use and studied profoundly.

Tarantola [12] used a generalized least-squares method to develop a theory for the nonlinear, multidimensional inversion of multioffset seismic reflection data in the acoustic approximation. Then, Tarantola [13] extended the method to the inversion of seismic data received on the surface of isotropic elastic media. Gauthier et al. [2] testified Tarantola's [12] theory by nonlinear inversion of theoretical multioffset seismic reflection data. Shortly after that, Tarantola [14] studied a strategy for nonlinear elastic inversion of seismic reflection data. Pan et al. [6] discussed the nonlinear inversion of plane-wave seismograms in stratified media by $p - \tau$ transform. In order to utilize more information in data including reflection on surface and transmission data of VSP, Mora [5] did some work. When only reflection data are used, the inversion mainly resolves sudden P - and S -wave impedance changes because these are what generate reflections. When transmission data are used in inversion, the large-scale P - and S -wave velocity variations, such as blockness are resoved. Sambridge et al. [8] improved the method for modifing models by subspace method, by which P -, S -wave impedance and density are obtained simultaneously. Zhou [17] extended the gradient nonlinear inversion technique proposed and developed by Tarantola and other authors to the inversion of multidimensional, multicomponent and multioffset seismograms received on the surface of arbitrary elastic anisotropic media.

The combined genetic and NAI uses advantages of NAI and GA and overcomes their shortcomings. Thus the combined inversion method is not only able to find global solution, but also to converge quickly. Porsani et al. [7] have undergone some studies on the combined genetic and linear inversion algorithm for isotropic media.

In this paper we will discuss nonlinear analytic inversion method for anisotropic media [17], and the combined genetic and nonlinear analytic inversion method [17]. About GA readers may refer literatures [3,9,10,11].

NONLINEAR ANALYTIC INVERSION METHOD FOR ANISOTROPIC MEDIA

Gradient Algorithm

The aim of seismic inversion is to determine distributions of density $\rho(\mathbf{x})$ and elastic coefficients $c_{ijkl}(\mathbf{x})(i,j,k,l=1,2,3)$ from displacement fields $u_{obs}(\mathbf{x}_r,t)$ received on the surface of or in the earth. Generally, $\rho(\mathbf{x})$ and $c_{ijkl}(\mathbf{x})$ are called model. $\mathbf{x}=(x_1, x_2,x_3)$ is a position of a point and $\mathbf{x}_r(r=1,2,....)$ a receiver position and t a time variable.

The inversion problem is to obtain the model $(\rho(\mathbf{x}), c_{ijkl}(\mathbf{x}))$ for which the misfit function or objective function

$$S = 1/2[(\mathbf{u} - \mathbf{u}_{obs})\,{}^t\mathbf{C}_d^{-1}(\mathbf{u} - \mathbf{u}_{obs}) + (\mathbf{m} - \mathbf{m}_0)\,{}^t\mathbf{C}_m^{-1}(\mathbf{m} - \mathbf{m}_0)] \qquad (1)$$

is a minimum. This criterior can be generalized to take into account the estimated observational errors and a priori information in the model space. In equation (1), u and u_{obs} are wave fields of predicted model m and observed wave fields respectively, m_0 is a priori model. C_d and C_m are covariance operators of observed data and priori model. The superscript t means transpose operation. Seismogram u and model m are related by

$$u = g(m), \qquad (2)$$

where g is a nonlinear operator describing the forward modeling.

Let m_n be a model in nth iteration, then we have

$$u_n = g(m_n), \qquad (3a)$$
$$S = \Delta u_n^t C_d^{-1} \Delta u_n + \Delta m^t C_m^{-1} \Delta m, \qquad (3b)$$
$$\Delta u_n = u_n - u_{obs}, \qquad (3c)$$
$$\Delta m = m_n - m_0, \qquad (3d)$$

and iterative formulae of conjugate gradient algorithms [15]

$$S(m_n + \delta m) = S(m_n) + \langle \hat{r}_n, \delta m \rangle + O(\|\delta m\|^2), \qquad (4a)$$
$$\hat{r}_n = G_n^t C_d^{-1} \Delta u_n + C_m^{-1} \Delta m = (\hat{r}_{\rho n}, \hat{r}_{cn}), \qquad (4b)$$
$$\hat{r}_{\rho n} = \delta\hat{\rho} + C_\rho^{-1}(\rho_n - \rho_0), \quad \hat{r}_{cn} = \delta\hat{c}_{ijkl} + C_c^{-1}(c_{ijkln} - c_{ijkl0}), \qquad (4c)$$
$$r_n = C_m \hat{r}_n, \qquad (4d)$$
$$\varphi_n = r_n + \alpha_n \varphi_{n-1}, \quad \varphi_1 = r_1, \qquad (4e)$$
$$\alpha_n = \frac{r_n^t(\hat{r}_n - \hat{r}_{n-1})}{r_n^t \hat{r}_n}, \qquad (4f)$$
$$m_{n+1} = m_n - \delta m, \quad \delta m = \eta_n \varphi_n, \qquad (4g)$$

where G_n is the Frechet derivative of g at point m_n of the model space, η_n is a constant, which can be calculated by different ways [5, 8].

The gradient $\hat{r}_n$ is an element of the dual of the model space, where dual and model space elements are related through the model covariance operator $C_m = \text{Diag}(C_\rho, C_c)$ by equation (4d). We will give the expressions of $\delta\hat{\rho}$ and $\delta\hat{c}_{ijkl}$ later.

Equations (4) are fundamental formulae of gradient inversion approach. When every variable in the equations is known, we are able to carry out the iterations. Of these variables the gradient vector is an critical one.

Objective function

In the context of least-squares, a weighting function is the integral kernal of the inverse of a covariance operator. If there are uncorrelate errors in the data set, depending on time or source and receiver positions, then we have [14]

$$\|u - u_{obs}\|^2 = \sum_s \sum_r \int_0^T dt \sum_{i=1}^3 \left[\frac{u^i(x_r, t; x_s) - u_{obs}^i(x_r, t; x_s)}{\sigma^2(t, x_r, x_s)}\right]^2 = \sum_s \sum_r \int_0^T dt \sum_{i=1}^3 [\delta d^i]^2. \qquad (5)$$

where σ^2 is the estimated (squared) error of data.

Elastodynamic Wave Equations for Anisotropic Media

Let $f_i(x, t; x_s)$ be body force component of sth shot, $T_i(x, t; x_s)$ stress component on the surface S, n_i normal vector component of S. The displacement corresponding to s^{th} shot is described by [15]

$$\rho(x)\frac{\partial^2 u^i}{\partial t^2}(x, t; x_s) - \frac{\partial}{\partial x_j}\left[c_{ijkl}(x)\frac{\partial u^k}{\partial x_l}(x, t; x_s)\right] = f_i(x, t; x_s),$$

$$n_j(x)c_{ijkl}(x)\frac{\partial u^k}{\partial x_l}(x, t; x_s) = T_i(x, t; x_s) \quad (x \in S),$$

$$u^i(x, 0; x_s) = 0,$$

$$\frac{\partial u^i}{\partial t} (x, 0; x_s) = 0. \tag{6}$$

The Gradient Vector for Anisotropic Media

The components of the gradient vector in equation (4c) are [17]

$$\delta\hat{\rho}(x) = -\sum_s \int_0^T dt \, \frac{\partial u_s^i}{\partial t} (x, t; x_s) \, \frac{\partial \psi_s^i}{\partial t} (x, t; x_s), \tag{7}$$

$$\delta\hat{c}_{ijkl}(x) = -\sum_s \int_0^T dt \, \frac{\partial \psi_s^k}{\partial x_l} (x, t; x_s) \, \frac{\partial u_s^i}{\partial x_j} (x, t; x_s), \tag{8}$$

where, T is the length of seismogram, ψ_s the back propagating field obeys

$$\rho_s(x) \frac{\partial^2 \psi_s^k}{\partial t^2} (x, t; x_s) - \frac{\partial}{\partial x_j} \Big[c_{ijkl}(x) \, \frac{\partial \psi_s^k}{\partial x_l} (x, t; x_s) \Big] = 0,$$

$$n_j(x) c_{ijkl}(x) \frac{\partial \psi_s^k}{\partial x_l} (x, t; x_s) = \sum_r \delta(x - x_r) \delta \hat{u}^i (x, t; x_s) \quad (x \in S),$$

$$\psi_s^i(x, T; x_s) = 0, \tag{9}$$

$$\frac{\partial \psi_s^k}{\partial t} (x, T; x_s) = 0,$$

where, $t \in [T, 0]$, $\psi_s(x, t; x_s)$ satisfies final condition.

COMBINED INVERSION METHOD

Advantages and shortcomings of GA and NAI

Genetic algorithm is a stochastic search technique, employing probabilistic mechanisms to solve complex optimization problems with multiple minima. The method starts with an ensemble of randomly selected models, requires very little priori information, and extensively and efficiently samples the most significant parts of model space. The global optimization method for seismic waveform inversion for anisotropic media is computationally expensive because the model space is very large. Though GA is based on the rules of natural selection and natural genetics, searches the most optimal parts of model space to find the global solution, it is a time consume method. As a result it is difficult to apply GA to inversion of practical data, particularly to large inversion problems.

The gradient information of objective function in NAI is the basis of modification of model and NAI will find a local minimum in the close neighborhood of a starting model. If the choice of the starting model is proper, i. e., it is just located in the valley whose minimum is the global solution, the iterations of NAI converge to the global solution. However, the probability of the starting model being near the global solution is very small, especially for the situation of no any priori information of model being known. But we should mension that NAI has high convergence rate.

It is meaningful to absorb the advantages of high convergence rate of NAI and of the ability to find global solution of GA, and to overcome their weaknesses of searching for local solution of NAI and of huge computational work of GA. The combined genetic and NAI is able to lead to this aim. In the combined inversion method, the function of GA is to provide models close to global solution and that of NAI to find the global solution as quickly as possible. Thus, the combined inversion method has the ability to find global solution, and converges faster than GA used alone.

Combining Scheme

In the course of optimization, GA continuously explores the whole model space. At the end of each GA iteration, an optimal model is obtained. According to the mathematical foundations of GA [3], the number of the optimal model in the next generation may increase, and at the same time there are new model created by the functions of crossover, selection and mutation. Of the population of the next generation the optimal model perhaps is the same as the one of previous generation, and maybe is a new one which is worse than the former. All this optimal models may be in the same valley of objective function as the previous one, or be in another valley. From the GA process of search optimal model we see that several times of GA iteration are needed to determine an extreme value. As a result, the randomness of GA leads that GA is time consume. However the randomness of GA guarantees to find the global solution.

If we take the optimal model of each GA iteration as the starting model of NAI, NAI can find a local minimum in the close neighborhood of the starting model. For NAI is a deterministic method, it modifies the model in the direction of the steepest descent of objective function. NAI finds the local minimum only after few times of iteration. Whether the local minimum is the global solution, NAI can not guarantee. Only does GA provide a starting model close to the global optimal solution, NAI may obtain the global solution.

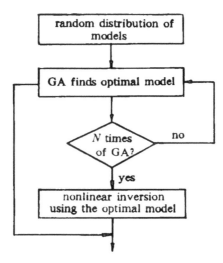

Figure 1. Diagram of combined inversion method. *N* is the iterative number of continuous GA iteration.

The match of GA and NAI in the combined inversion method is an important problem. After NAI gets the local minimum close to the starting model provided by GA, the local minimum may be the same one in the next several generations because of the randomness of GA. If NAI is undergone at the end of each GA iteration, the results of combined inversion are the same. Obviously, some computation of NAI here is not necessary. Therefore, we propose a combining strategy that after several consecutive GA iterations, one time of NAI is done. Then, the result of NAI is regarded as one model of population of next GA iteration. Fig. 1 shows the diagram of combined inversion method.

EXAMPLES

In order to testify the advantages of the combined inversion method, we use examples of inversion of parameters of an one-dimensional multi-layered transversely isotropic medium model.

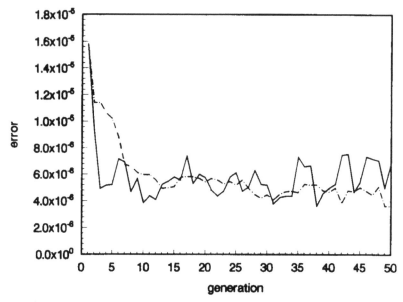

Figure 2. Objective functions of NAI (solid) and GA (dash) versus the number of generations. One time of NAI is applied at the end of each GA iteration in CI.

Fig. 2 is the curves of objective function values versus the number of generations. In this example NAI is done at the end of each GA iteration. The objective function drops sharply at the first several few iterations, particularly at the first three times. The little error the combined method achieves at the end of 10th iteration, GA reaches the value at 42nd generation. The curve of CI leaps severer than GA's. This can be interpreted as following. The model derived by NAI acts as a parent in the next generation. This model is usually replaced by new model because of the randomness of GA. The new and old models may be located in different valleys of objective function, so the solution of NAI with the new starting model may be different from the previous.

Although the curve of combined method is oscillating, there are some plateaus (i. e., parts of curve with unchangable values of objective function). This phenomenon indicates that the optimal models in these generations are the same. NAI in the combined inversion method can not offer any improvement. Therefore, NAI in this situation is not necessary, otherwise combined inversion method becomes expensive.

Fig. 3 is another example of five times of GA iteration plus one time of NAI. In the combined inversion method, GA is a main procedure. The values of objective function of combined inversion method are evidently less than these of GA at the same generations. The error of CI decreases suddenly at the end of 5th iteration. GA reaches the error at the end of 10th iteration of CI at 42nd generation, and GA never reaches the smallest error of CI. The curve of CI is stable in the entire course of iteration since GA is a dominant procedure.

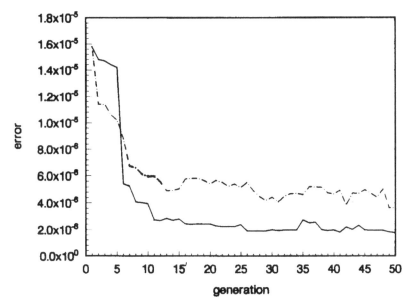

Figure 3. Objective functions of NAI (solid) and GA (dash) versus the number of generations. One time of NAI is applied at the end of per five GA iterations in CI.

CONCLUSIONS

Combined nonlinear inversion method, which makes use of goodnesses and overcomes weaknesses of GA and NAI, is a useful method for finding global solution with high convergence rate.

The combining scheme of GA and NAI in CI is important. From theoretical examples we conclude that the effect of CI with five GA iterations plus one time of NAI is much better than that of CI with one time of GA and one time of NAI. But, how many times of continuous GA iteration in CI is the best remains a problem to be studied.

The function of GA in CI is to provide models close to global solution. The rate of combined inversion is mainly determined by GA. We think that the theory of uniform design may be applied to GA in order to speed the random exploration.

Like GA, other random walk methods can be used to form combined inversion methods with NAI too.

REFERENCES

1. E. Berg. Simple convergent genetic algorithm for inversion of multiparameter data. *SEG*60 *Expanded Abstracts*, II, 1126-1128 (1990).
2. O. Gauthier, J. Virieux and A. Tarantola. Two-dimensional nonlinear inversion of seismic waveforms: Numerical results. *Geophysics*, **51**, 1387-1403 (1986).
3. D. E. Goldberg. *Genetic Algorithms in Search, Optimiztion, and Machine Learning*. Addison-Wesley, Reading, MA (1989).
4. J. H. Holland. *Adaptation in Natural and Artifical Systems*, The University of Michigan Press, Ann Arbor (1975).
5. P. Mora. 2D elastic inversion of multi-offset seismic data. *Geophysics*, **52**, 2031-2050 (1988).
6. G. S. Pan, R. A. Phinney and R. I. Odom. Full-waveform inversion of plane-wave seismograms in stratified acoustic media: Theory and feasibility. *Geophysics*, **53**, 21-31 (1988).
7. M. J. Porsani, P. L. Stoffa, M. K. Sen, et al. A combined Genetic and linear inversion algorithm for seismic waveform inversion. *SEG*63 *Expanded Abstracts*, 692-695 (1993).
8. M. S. Sambridge, A. Tatantola and Kennet. An alternative strategy for nonlinear inversion of seismic waveforms. *Geophysical Prospecting*, **39**, 723-736 (1991).
9. M. Sambridge and G. Drijkoningen. Genetic algorithms in seismic waveform inversion. *Geophys. J. Int.* **109**, 323-342 (1992).
10. M. K. Sen and P. L. Stoffa. Rapid sampling of model space using genetic algorithms: examples from seismic waveform inversion. *Geophys. J. Int.* **109**, 323-342 (1992).
11. P. L. Stoffa and M. K. Sen. Nonlinear multiparametre optimization using genetic algorithms: Inversion of plane-wave seismograms. *Geophysics*, **56**, 1794-1810 (1991).
12. A. Tarantola. Inversion of seismic reflection data in the acoustic approximation. *Geophysics*, **49**, 1259-1266 (1984).
13. A. Tarantola. The seismic reflection inverse problem, in Santosa, F., Pao, Y.-H., system, W. W., and Holland, C. Eds, *Inverse problems of acoustic and elastic waves*. Soc. Industr. Appl. Math., 104-181 (1984).
14. A. Tarantola. A strategy for nonlinear elastic inversion of seismic reflection data. *Geophysics*, **51**, 1893-1903 (1986).
15. A. Tarantola. *Inverse problem theory: Methods for data fitting and model parameter estimation*. Elsevier Science Publ. Co. Inc. (1987).
16. H. Zhou and Q. He. Genetic algorithms in anisotropic parameter inversion, *Journal of Changchun university of earth sciences*, Sup. 1, 62-67 (1995).
17. H. Zhou. *A study on modeling and nonlinear inversion of elastic wave equations in anisotropic media: Ph. D. thesis*, Changchun Univ. of Earth sciences (1995).

Proc. 30ʰ Int'l. Geol. Congr., Vol. 4, pp. 53-63
Hong Dawei (Ed)
© VSP 1997

A Finite-Difference Scheme and Its Stability Criterion for Elastic-Wave Equations in Anisotropic Media

DINGHUI YANG
Department of Geosciences, University of Petroleum, Beijing 100083, P. R. China
JIWEN TENG and ZHONGJIE ZHANG
Institute of Geophysics, Academia Sinica, Beijing 100101, P. R. China

Abstract

A fast and suitable algorithm for modeling seismic wave propagation in anisotropic heterogeneous media is of considerable value in current exploration seismology. The stability analysis of an algorithm is necessary for improving computational rate. In this paper, we first express the elastic wave equations by use of matrices and vectors; then we suggest a fast and small storage finite-difference scheme for 2-dimensional and 3-dimensional elastic wave equations in general anisotropic media. Second, we systematically study the stability criterion of the finite-difference method for the 2-D equations of wave-motion in homogeneous and inhomogeneous media. Furthermore, we give several stability criteria for some special elastic media such as the isotropic medium and transversely isotropic medium in detail. Finally, this paper presents the stability criterion for the 3-D finite-difference scheme, too.

Keywords: Elastic-wave equations, Finite-difference, Stability, Anisotropic media.

INTRODUCTION

The numerical simulation of seismic propagation has important significance in Geosciences. Among the various techniques available for seismic modelling in an anisotropic medium, the most popular numerical technique is reflectivity method [3, 4], which is based on Kennett's work [11, 12]. Ray tracing method, which is based on the asymptotic solution of the eikonal equation, is another technique used for modelling anisotropic media [5, 6]. Kosloff et al. [13] used Fourier methods to compute the spatial derivatives and rapid expansion method as a time integration algorithm to model a 3-D anisotropic medium. Chen [7] used finite element methods to model an anisotropic heterogeneous medium. Finite-difference methods, although widely applied to elastic modelling in isotropic media, are not common in modelling anisotropic media. Tsingas et al. [16] present a modelling algorithm that uses finite-difference operators. Their algorithm is based on MacCormack-type splitting scheme [2] for the solution of differential equations that describe wave propagation in transversely isotropic media. Faria et al. [8] used finite-difference algorithms to model a 2-D transversely isotropic medium. This method is based on the staggered grid scheme [17]. Recently, Igel et al. [9] present a finite-difference algorithm for modelling general anisotropic media. Their method is based on the convolution algorithm.

Rapidity and less storage etc. are advantages of the finite–difference method. With the requirements of large scale wave–fields modelling and the developments of massively parallel calculations for seismic modelling in general anisotropic media, finite–difference schemes for seismic modelling in complicated media or high dimension (2–D or 3–D) models were inconceivable. Pseudospectral methods are attractive since the space operators are exact up to the Nyquist frequency. However, they require Fourier transforms of the fields to be made, which are computationally expensive for 3–D anisotropic models. Moreover, taking the Fourier transform means that each point interacts with every other point. In some sense, this is unphysical as interaction in dynamic elasticity is of a local nature. Thus it is necessary to take into account local operators as we design a finite–difference scheme for seismic modelling. On the other hand, considering local difference operators is also important to accelerate the massive parallelism since nearest neighbor communication is extremely fast, and large anisotropic models are feasible because of the intrinsic parallelism of the coservation equations.

Considering the reasons stated above, we present a fast and small storage finite–difference method for modelling seismic wave propagation in general anisotropic media since only three nearest grid points are considered in the scheme. The method is an extension of the algorithm [10,18] to general anisotropic case.

Usually, time–marching algorithm is used in modelling seismic propagation. In order to keep an algorithm stable, the time increment in the time–marching algorithm is limited by stability conditions. Provided a suitable time increment is chosen, not only stable calculation can be kept, but iterative speed can be improved while the spatial increment is constants. Otherwise, unphysical numerical oscillation can produce, even result in mistaken results as an unsuitable time increment is chosen. choosing a good time increment is decided by the finite–difference and the parameters or velocities which describe a medium characteristics. i.e., our choosing a good time increment is decided by the stability condition of the finite–difference scheme. Therefore, stability of finite–difference schemes is very important in numerically modelling seismic wave–fields. Although our work is an extension of the algorithms [10,18], they do not present systematic and detailed stability results of the finite–difference method in general anisotropic cases. This work is motivated by the need to find the stability criteria for the general anisotropic case. Furthermore, some more detailed stability criteria for the finite–difference scheme in some special media are provided. Our result in the isotropic case is consistent with that of Aboudi [1].

FINITE–DIFFERENCE SCHEME

3–dimensional anisotropic medium

The motional equilibrum equations in an anisotropic medium can be written as follows:

$$\frac{\partial \sigma_{xx}}{\partial x} + \frac{\partial \sigma_{xy}}{\partial y} + \frac{\partial \sigma_{xz}}{\partial z} + f_x = \rho \frac{\partial^2 u_x}{\partial t^2}, \tag{1a}$$

$$\frac{\partial \sigma_{xy}}{\partial x} + \frac{\partial \sigma_{yy}}{\partial y} + \frac{\partial \sigma_{yz}}{\partial z} + f_y = \rho \frac{\partial^2 u_y}{\partial t^2}, \text{ and} \tag{1b}$$

$$\frac{\partial \sigma_{xz}}{\partial x} + \frac{\partial \sigma_{yz}}{\partial y} + \frac{\partial \sigma_{zz}}{\partial z} + f_z = \rho \frac{\partial^2 u_z}{\partial t^2},$$ (1c)

where ρ is the density of the medium; f_x, f_y and f_z denote the components of force source in the x, y and z direction, respectively; and u_x, u_y and u_z denote displacement components in the x, y, and z direction, respectively.

The relationship between stress and strain tensor is $\sigma = \bar{C}\varepsilon$, where,

the elastic parameter matrix $\bar{C} = (c_{i,j})_{6\times6}$, and $c_{i,j} = c_{j,i}$, i, j= 1, 2, ..., 6;

the stress matrix $\sigma = (\sigma_{xx},\ \sigma_{yy},\ \sigma_{zz},\ \sigma_{yz},\ \sigma_{xz},\ \sigma_{xy})^T$;

the strain matrix $\varepsilon = (\varepsilon_{xx},\ \varepsilon_{yy},\ \varepsilon_{zz},\ \varepsilon_{yz},\ \varepsilon_{xz},\ \varepsilon_{xy})^T$,

and the strain–displacement relation

$$\varepsilon_{xx} = \frac{\partial u_x}{\partial x},\ \varepsilon_{yy} = \frac{\partial u_y}{\partial y},\ \varepsilon_{zz} = \frac{\partial u_z}{\partial z},\ \varepsilon_{xy} = \frac{\partial u_x}{\partial y} + \frac{\partial u_y}{\partial x},$$

$$\varepsilon_{xz} = \frac{\partial u_x}{\partial z} + \frac{\partial u_z}{\partial x},\ \varepsilon_{yz} = \frac{\partial u_y}{\partial z} + \frac{\partial u_z}{\partial y}.$$

Let

$$A = \begin{bmatrix} c_{11} & c_{16} & c_{15} \\ c_{16} & c_{66} & c_{56} \\ c_{15} & c_{56} & c_{55} \end{bmatrix},\ B = \begin{bmatrix} c_{16} & c_{12} & c_{14} \\ c_{66} & c_{26} & c_{46} \\ c_{56} & c_{25} & c_{45} \end{bmatrix},\ C = \begin{bmatrix} c_{15} & c_{14} & c_{13} \\ c_{56} & c_{46} & c_{36} \\ c_{55} & c_{45} & c_{35} \end{bmatrix},$$

$$D = \begin{bmatrix} c_{16} & c_{66} & c_{56} \\ c_{12} & c_{26} & c_{25} \\ c_{14} & c_{46} & c_{45} \end{bmatrix},\ E = \begin{bmatrix} c_{66} & c_{26} & c_{46} \\ c_{26} & c_{22} & c_{24} \\ c_{46} & c_{24} & c_{44} \end{bmatrix},\ F = \begin{bmatrix} c_{56} & c_{46} & c_{36} \\ c_{25} & c_{24} & c_{23} \\ c_{45} & c_{44} & c_{34} \end{bmatrix},$$

$$G = \begin{bmatrix} c_{15} & c_{56} & c_{55} \\ c_{14} & c_{46} & c_{45} \\ c_{13} & c_{36} & c_{35} \end{bmatrix},\ H = \begin{bmatrix} c_{56} & c_{25} & c_{45} \\ c_{46} & c_{24} & c_{44} \\ c_{36} & c_{23} & c_{34} \end{bmatrix},\ Q = \begin{bmatrix} c_{55} & c_{45} & c_{35} \\ c_{45} & c_{44} & c_{34} \\ c_{35} & c_{34} & c_{33} \end{bmatrix},$$

$$U = (u_x,\ u_y,\ u_z)^T,\ \bar{F} = (f_x,\ f_y,\ f_z)^T,$$

then equation (1) can be written as follows:

$$\rho \frac{\partial^2 U}{\partial t^2} = (\frac{\partial}{\partial x}(A\frac{\partial}{\partial x} + B\frac{\partial}{\partial y} + C\frac{\partial}{\partial z}) + \frac{\partial}{\partial y}(D\frac{\partial}{\partial x} + E\frac{\partial}{\partial y} + F\frac{\partial}{\partial z})$$

$$+ \frac{\partial}{\partial z}(G\frac{\partial}{\partial x} + H\frac{\partial}{\partial y} + Q\frac{\partial}{\partial z}))U + \bar{F}.$$ (2)

Obviously, A, E, Q, B+D, C+G and F+H are real symmetric matrices.

Adopting finite–difference to approximate equation (2) without the source term F, we can obtain the following finite–difference scheme:

$$U_{j,l,k}^{n+1} = 2U_{j,l,k}^{n} - U_{j,l,k}^{n-1} + (\frac{\Delta t}{\Delta x})^2 \frac{1}{\rho_{j,l,k}} [A_{j+\frac{1}{2},l,k}(U_{j+1,l,k}^{n} - U_{j,l,k}^{n})$$

$$- A_{j-\frac{1}{2},l,k}(U_{j,l,k}^{n} - U_{j-1,l,k}^{n})] + (\frac{\Delta t}{\Delta y})^2 \frac{1}{\rho_{j,l,k}} [E_{j,l+\frac{1}{2},k}(U_{j,l+1,k}^{n} - U_{j,l,k}^{n})$$

$$- E_{j,l-\frac{1}{2},k}(U_{j,l,k}^{n} - U_{j,l-1,k}^{n})] + (\frac{\Delta t}{\Delta z})^2 \frac{1}{\rho_{j,l,k}} [Q_{j,l,k+\frac{1}{2}}(U_{j,l,k+1}^{n} - U_{j,l,k}^{n})$$

$$- Q_{j,l,k-\frac{1}{2}}(U_{j,l,k}^{n} - U_{j,l,k-1}^{n})] + \frac{(\Delta t)^2}{4\Delta x \Delta y \rho_{j,l,k}} [B_{j+1,l,k}(U_{j+1,l+1,k}^{n}$$

$$- U_{j+1,l-1,k}^{n}) - B_{j-1,l,k}(U_{j-1,l+1,k}^{n} - U_{j-1,l-1,k}^{n})]$$

$$+ \frac{(\Delta t)^2}{4\Delta x \Delta z \rho_{j,l,k}} [C_{j+1,l,k}(U_{j+1,l,k+1}^{n} - U_{j+1,l,k-1}^{n}) - C_{j-1,l,k}(U_{j-1,l,k+1}^{n}$$

$$- U_{j-1,l,k-1}^{n})] + \frac{(\Delta t)^2}{4\Delta x \Delta y \rho_{j,l,k}} [D_{j,l+1,k}(U_{j+1,l+1,k}^{n} - U_{j-1,l+1,k}^{n})$$

$$- D_{j,l-1,k}(U_{j+1,l-1,k}^{n} - U_{j-1,l-1,k}^{n})] + \frac{(\Delta t)^2}{4\Delta y \Delta z \rho_{j,l,k}} [F_{j,l+1,k}(U_{j,l+1,k+1}^{n}$$

$$- U_{j,l+1,k-1}^{n}) - F_{j,l-1,k}(U_{j,l-1,k+1}^{n} - U_{j,l-1,k-1}^{n}) + H_{j,l,k+1}(U_{j,l+1,k+1}^{n}$$

$$- U_{j,l-1,k+1}^{n}) - H_{j,l,k-1}(U_{j,l+1,k-1}^{n} - U_{j,l-1,k-1}^{n})]$$

$$+ \frac{(\Delta t)^2}{4\Delta x \Delta z \rho_{j,l,k}} [G_{j,l,k+1}(U_{j+1,l,k+1}^{n} - U_{j-1,l,k+1}^{n}) - G_{j,l,k-1}(U_{j+1,l,k-1}^{n}$$

$$- U_{j-1,l,k-1}^{n})]. \tag{3}$$

where, Δx, Δy and Δz denote spatial increments in the x, y and z direction, respectively; Δt denotes the time-step size, and

$$A_{j+\frac{1}{2},l,k} = \frac{1}{2}(A_{j+1,l,k} + A_{j,l,k}) \ , \ E_{j,l+\frac{1}{2},k} = \frac{1}{2}(E_{j,l+1,k} + E_{j,l,k}) \ ,$$

$$Q_{j,l,k+\frac{1}{2}} = \frac{1}{2}(Q_{j,l,k+1} + Q_{j,l,k}).$$

2-Dimensional anisotropic medium

Similarly, we can express the wave-equations in 2-D anisotropic medium as follows:

$$\rho \frac{\partial^2 U}{\partial t^2} = (\frac{\partial}{\partial x}(A\frac{\partial}{\partial x} + C\frac{\partial}{\partial z}) + \frac{\partial}{\partial z}(G\frac{\partial}{\partial x} + Q\frac{\partial}{\partial z}))U + \bar{F} \ , \tag{4}$$

and have the following difference scheme:

$$U_{j,l}^{n+1} = 2U_{j,l}^{n} - U_{j,l}^{n-1} + (\frac{\Delta t}{\Delta x})^2 \frac{1}{\rho_{j,l}} [A_{j+\frac{1}{2},l}(U_{j+1,l}^{n} - U_{j,l}^{n}) - A_{j-\frac{1}{2},l}(U_{j,l}^{n} - U_{j-1,l}^{n})]$$

$$+ (\frac{\Delta t}{\Delta z})^2 \frac{1}{\rho_{j,l}} [Q_{j,l+\frac{1}{2}}(U_{j,l+1}^{n} - U_{j,l}^{n}) - Q_{j,l-\frac{1}{2}}(U_{j,l}^{n} - U_{j,l-1}^{n})]$$

$$+ \frac{(\Delta t)^2}{4\Delta x \Delta z} \frac{1}{\rho_{\lambda, l}} [C_{j+1, l}(U^n_{j+1, l+1} - U^n_{j+1, l-1}) - C_{j-1, l}(U^n_{j-1, l+1} - U^n_{j-1, l-1})$$

$$+ G_{\lambda, l+1}(U^n_{j+1, l+1} - U^n_{j-1, l+1}) - G_{\lambda, l-1}(U^n_{j+1, l-1} - U^n_{j-1, l-1})]. \tag{5}$$

THE STABILITY CRITERION

The stability criterion for the 2–D finite–difference scheme in homogeneous media

The finite–difference scheme (5) in a homogeneous medium case can be simply written as follows:

$$U^{n+1}_{\lambda, l} = 2U^n_{\lambda, l} - U^{n-1}_{\lambda, l} + (\frac{\Delta t}{\Delta x})^2 \frac{1}{\rho} A(U^n_{j+1, l} - 2U^n_{\lambda, l} + U^n_{j-1, l}) + \frac{(\Delta t)^2}{4\rho \Delta x \Delta z}(C$$

$$+ G)(U^n_{j+1, l+1} - U^n_{j+1, l-1} - U^n_{j-1, l+1} + U^n_{j-1, l-1}) + \frac{(\Delta t)^2}{\rho(\Delta z)^2} Q(U^n_{\lambda, l+1}$$

$$- 2U^n_{\lambda, l} + U^n_{\lambda, l-1}) . \tag{6}$$

According to Richtmyer and Morton's theoretical stability analysis [16], we set

$$U^n_{\lambda, l} = U_0 e^{i(\omega \lambda \Delta x + \gamma l \Delta z)} \lambda^n ,$$

where U_0 is a constant vector. System (6) reduces to

$$(\lambda^2 I - 2P\lambda + I)U_0 = 0.$$

where I denotes an identity matrix,

$$P = I - 2[k_1 \sin^2 \alpha A + k_2 \sin^2 \beta Q + \frac{1}{4} k_3 \sin 2\alpha \sin 2\beta (C + G)],$$

with $k_1 = \frac{(\Delta t)^2}{\rho(\Delta x)^2}$, $k_2 = \frac{(\Delta t)^2}{\rho(\Delta z)^2}$, $k_3 = \frac{(\Delta t)^2}{\rho \Delta x \Delta z}$, $\alpha = \frac{1}{2}\omega \Delta x$, $\beta = \frac{1}{2}\gamma \Delta z$.

Equation stated above is satisfied if the determinant of coefficients vanishes:

$$|\lambda^2 I - 2P\lambda + I| = 0 . \tag{7}$$

Theorem 1. The stability criterion for the finite–difference scheme (6) is

$$\|A\|(\frac{\Delta t}{\Delta x})^2 + \|Q\|(\frac{\Delta t}{\Delta z})^2 \leq \rho, \text{ if } \|C + G\| \leq 2\sqrt{\|A\|\|Q\|},$$

$$\max(F(\alpha, \beta), \|A\|(\frac{\Delta t}{\Delta x})^2 + \|Q\|(\frac{\Delta t}{\Delta z})^2) \leq \rho, \text{ if } \|C + G\| > 2\sqrt{\|A\|\|Q\|}.$$

where the function $F(\alpha, \beta)$, α, β are

$$F(\alpha, \beta) = (\frac{\Delta t}{\Delta x})^2 \|A\|\sin^2 \alpha + (\frac{\Delta t}{\Delta z})^2 \|Q\|\sin^2 \beta + \frac{(\Delta t)^2}{4\Delta x \Delta z} \|C + G\|\sin 2\alpha \sin 2\beta,$$

$$\sin 2\alpha = \frac{\|Q\|}{\|C + G\|^2} (\frac{\|C + G\|^4 - 16\|A\|^2\|Q\|^2}{k^4\|A\|^2 + \|Q\|^2})^{1/2},$$

$$\sin 2\beta = \frac{k\|C + G\|\sin 2\alpha}{\sqrt{4\|Q\|^2 + k^2\|C + G\|^2 \sin^2 2\alpha}}.$$

where $k = \Delta z / \Delta x$.

Proof. According to the stability of the difference scheme (6), we have $\|\lambda\| \leqslant 1$ in equation (7).

In terms of equation (7) and lemma 2 (see Appendix), we can obtain the following inequality:

$$\|P\| \leqslant 1. \tag{8}$$

Because the matrices A, Q and C+G are symmetric, the following matrix

$$2[k_1 A \sin^2 \alpha + k_2 Q \sin^2 \beta + \frac{1}{4} k_3 (C + G)\sin 2\alpha \sin 2\beta]$$

is also symmetric. Then, in terms of lemma 3 (see Appendix) and inequality (8), we have

$$\|k_1 A \sin^2 \alpha + k_2 Q \sin^2 \beta + \frac{1}{4} k_3 (C + G)\sin 2\alpha \sin 2\beta\| \leqslant 1. \tag{9}$$

Because the elements of the matrices A, Q and C+G are non−negative, let $0 \leqslant \alpha, \beta \leqslant \frac{\pi}{2}$, if we want to make (9) correct, we only require

$$k_1 \|A\| \sin^2 \alpha + k_2 \|Q\| \sin^2 \beta + \frac{1}{4} k_3 \|C + G\| \sin 2\alpha \sin 2\beta \leqslant 1.$$

Let

$$f(\alpha, \beta) = k_1 \|A\| \sin^2 \alpha + k_2 \|Q\| \sin^2 \beta + \frac{1}{4} k_3 \|C + G\| \sin 2\alpha \sin 2\beta. \tag{10}$$

Set the partial derivatives

$$f_\alpha \equiv k_1 \|A\| \sin 2\alpha + \frac{1}{2} k_3 \|C + G\| \cos 2\alpha \sin 2\beta = 0, \tag{11}$$

$$f_\beta \equiv k_2 \|Q\| \sin 2\beta + \frac{1}{2} k_3 \|C + G\| \sin 2\alpha \cos 2\beta = 0. \tag{12}$$

According to the original characteristics of wave−motion equations, we know
$$\| A \| > 0, \quad \| Q \| > 0, \quad \| C+G \| > 0.$$

So, both $(0, 0)$ and $(\frac{\pi}{2}, \frac{\pi}{2})$ can be extreme points of the function $f(\alpha, \beta)$. Obviously,

$$f(0, 0) = 0,$$

$$f(\frac{\pi}{2}, \frac{\pi}{2}) = k_1 \|A\| + k_2 \|Q\|.$$

Next, we discuss the cases: $(\alpha, \beta) \neq (0, 0)$ or $(\frac{\pi}{2}, \frac{\pi}{2})$.

According to (11)−(12), if $\|C + G\| > 2\sqrt{\|A\| \|Q\|}$, then

$$\sin 2\alpha = \frac{\|Q\|}{\|C + G\|^2} \left(\frac{\|C + G\|^4 - 16 \|A\|^2 \|Q\|^2}{k^4 \|A\|^2 + \|Q\|^2} \right)^{\frac{1}{2}}, \tag{13}$$

$$\sin 2\beta = \frac{k \|C + G\| \sin 2\alpha}{\sqrt{4 \|Q\|^2 + k^2 \|C + G\|^2 \sin^2 2\alpha}}. \tag{14}$$

where α, β can be the maximal value point of $f(\alpha, \beta)$, so the stability criterion is

$$\max(F(\alpha, \beta), \rho f(\frac{\pi}{2}, \frac{\pi}{2})) \leqslant \rho.$$

Otherwise, $(\alpha, \beta) = (\frac{\pi}{2}, \frac{\pi}{2})$ is the maximum value point of the function $f(\alpha, \beta)$, and the stability criterion is

$$\|A\|(\frac{\Delta t}{\Delta x})^2 + \|Q\|(\frac{\Delta t}{\Delta z})^2 \leqslant \rho.$$

In particular, when $\Delta x = \Delta z$, we can obtain a simple stability criterion as follows:

$$\frac{\Delta t}{\Delta x} \leqslant \sqrt{\frac{\rho}{\|A\| + \|Q\|}}, \quad \text{if } \|C + G\| \leqslant 2\sqrt{\|A\|\|Q\|};$$

$$\frac{\Delta t}{\Delta x} \leqslant \left(\frac{\rho}{\max(\|A\| + \|Q\|, \ \|A\|\sin^2\alpha + \|Q\|\sin^2\beta + \frac{1}{4}\|(C + G)\|\sin2\alpha\sin2\beta)}\right)^{\frac{1}{2}},$$

if $\|C + G\| > 2\sqrt{\|A\|\|Q\|}$,

where α, β are defined by (13) and (14), and $k = 1$.

The stability criterion for inhomogeneous media

It is difficult to estimate the stability criterion for the difference scheme (5) in inhomogeneous media. However, according to the numerical method theory [15] of partial differential equations, we can adopt the "freezing coefficients" method [14] to analyse its stability. Furthermore, we give an approximate formula of the stability criterion for the difference scheme (5) in an inhomogeneous medium.

In fact, generally, we can deal approximately with a small computional area as a homogeneous medium if the functions of the medium parameters are continuous and bounded; then the difference scheme (5) can degenerate into the form (6), and we can obtain a stability criterion like that of the homogeneous medium in a small area. Furthermore, according to the bounded continuous character of the medium parameter founctions, we can obtain the stability criterion for the difference scheme (5) in an inhomogeneous medium as follows:

If $\max\limits_{i,j}[\|C_{i,j} + G_{i,j}\| - 2\sqrt{\|A_{i,j}\|\|Q_{i,j}\|}] \leqslant 0$, then

$$\max\limits_{i,j}\frac{\|A_{i,j}\|}{\rho_{i,j}}(\frac{\Delta t}{\Delta x})^2 + \max\limits_{i,j}\frac{\|Q_{i,j}\|}{\rho_{i,j}}(\frac{\Delta t}{\Delta z})^2 \leqslant 1.$$

Otherwise,

$$\max(H(\alpha, \beta), \ \max\limits_{i,j}\frac{\|A_{i,j}\|}{\rho_{i,j}}(\frac{\Delta t}{\Delta x})^2 + \max\limits_{i,j}\frac{\|Q_{i,j}\|}{\rho_{i,j}}(\frac{\Delta t}{\Delta z})^2) \leqslant 1,$$

where $H(\alpha, \beta)$, α, β are

$$H(\alpha, \beta) = \max\limits_{i,j}[\frac{\|A_{i,j}\|}{\rho_{i,j}}(\frac{\Delta t}{\Delta x})^2\sin^2\alpha + \frac{\|Q_{i,j}\|}{\rho_{i,j}}(\frac{\Delta t}{\Delta z})^2\sin^2\beta$$

$$+ \frac{(\Delta t)^2}{4\Delta x\Delta z}\frac{\|C_{i,j} + G_{i,j}\|}{\rho_{i,j}}\sin2\alpha\sin2\beta],$$

$$\sin2\alpha = \frac{\|Q_{i,j}\|}{\|C_{i,j} + G_{i,j}\|}\left(\frac{\|C_{i,j} + G_{i,j}\|^4 - 16\|A_{i,j}\|^2\|Q_{i,j}\|^2}{k^4\|A_{i,j}\|^2 + \|Q_{i,j}\|^2}\right)^{\frac{1}{2}},$$

$$\sin2\beta = \frac{k\|C_{i,j} + G_{i,j}\|\sin2\alpha}{\sqrt{4\|Q_{i,j}\|^2 + k^2\|C_{i,j} + G_{i,j}\|^2\sin^2 2\alpha}}.$$

The stability criterion for the difference scheme in some special media
Obviously, computing the maxium moduli of the matrices **A**, **Q** and **C+G** in the difference scheme (6), we can obtain the following stability criteria in isotropic and transversely isotropic media, respectively.

Isotropic media

$$(\frac{\triangle t}{\triangle x})^2(\lambda + 2\mu) + (\frac{\triangle t}{\triangle z})^2(\lambda + 2\mu) \leqslant \rho,$$

or

$$\alpha^2[(\frac{\triangle t}{\triangle x})^2 + (\frac{\triangle t}{\triangle z})^2] \leqslant 1, \tag{15}$$

where λ, μ are Lame parameters, $\alpha = \sqrt{\dfrac{\lambda + 2\mu}{\rho}}$ is the velocity of P—wave.

Obviously, the stability criterion (15) is the same as that of Aboudi [1].

Transversely isotropic media

If $F + L \leqslant 2\sqrt{\max(A, N, L)\max(L, C)}$, then

$$(\frac{\triangle t}{\triangle x})^2\max(A, N, L) + (\frac{\triangle t}{\triangle z})^2\max(L, C) \leqslant \rho,$$

otherwise,

$$\max(f1(\alpha, \beta), (\frac{\triangle t}{\triangle x})^2\max(A, N, L) + (\frac{\triangle t}{\triangle z})^2\max(L, C)) < \rho,$$

where,

$$f1(\alpha, \beta) = (\frac{\triangle t}{\triangle x})^2\max(A, N, L)\sin^2\alpha + (\frac{\triangle t}{\triangle z})^2\max(L, C)\sin^2\beta$$

$$+ \frac{(F + L)(\triangle t)^2}{4\triangle x\triangle z}\sin2\alpha\sin2\beta,$$

$$\sin2\alpha = \frac{\max(L, C)}{(F + L)^2}\left\{\frac{(F + L)^4 - 16[\max(A, N, L)\max(L, C)]^2}{k^4[\max(A, N, L)]^2 + [\max(L, C)]^2}\right\}^{\frac{1}{2}},$$

$$\sin2\beta = k(F + L)\sin2\alpha\{4[\max(L, C)]^2 + k^2(F + L)^2\sin^2 2\alpha\}^{-\frac{1}{2}},$$

where, A, N, L, F, C are the elastic constants.

Similarly, we can obtain similar stability criteria for an inhomogeneous medium and other special anisotropic media (e.g., the cubic anisotropic medium, perpendicular anisotropic medium etc.).

The stability criterion for the finite—difference in 3—D anisotropic media
As the analyses for the 2—D case before, we can obtain the following stability criterion:
Theorem 2. If

$$\max[k_1 \cdot \|A\| + k_2 \cdot \|E\| + k_3 \cdot \|Q\|, f2(\theta_1, \theta_2, \theta_3)] \leqslant 1,$$

then, the difference scheme (3) in homogeneous media case is stable. Where $\dfrac{\pi}{2}$ $\leqslant \theta_1, \theta_2, \theta_3 \leqslant \pi$; the function $f2(\theta_1, \theta_2, \theta_3)$ are defined in the following form:

$$f2(\theta_1, \theta_2, \theta_3) = k_1 \cdot \sin^2\frac{\theta_1}{2} \cdot \|A\| + k_2 \cdot \sin^2\frac{\theta_2}{2} \cdot \|E\| + k_3 \cdot \sin^2\frac{\theta_3}{2} \cdot \|Q\|$$

$$+ \frac{1}{4}k_4 \cdot \sin\theta_1 \cdot \sin\theta_2 \cdot \|B + D\| + \frac{1}{4}k_5 \cdot \sin\theta_1 \cdot \sin\theta_3 \cdot \|C + G\|$$

$$+ \frac{1}{4}k_6 \cdot \sin\theta_2 \cdot \sin\theta_3 \cdot \|F + G\|,$$

where

$$k_1 = \frac{1}{\rho} \cdot (\frac{\Delta t}{\Delta x})^2, \quad k_2 = \frac{1}{\rho} \cdot (\frac{\Delta t}{\Delta y})^2, \quad k_3 = \frac{1}{\rho} \cdot (\frac{\Delta t}{\Delta z})^2, \quad k_4 = \frac{1}{\rho} \cdot \frac{(\Delta t)^2}{\Delta x \cdot \Delta y},$$

$$k_5 = \frac{1}{\rho} \cdot \frac{(\Delta t)^2}{\Delta x \cdot \Delta z}, \quad k_6 = \frac{1}{\rho} \cdot \frac{(\Delta t)^2}{\Delta z \cdot \Delta y},$$

and θ_1, θ_2, θ_3 satisfy the following equalities:

$$tg\theta_1 = -\frac{1}{a} \cdot (g \cdot \sin\theta_2 + e \cdot \sin\theta_3),$$

$$tg\theta_2 = -\frac{1}{b} \cdot (g \cdot \sin\theta_1 + f \cdot \sin\theta_3), \text{ and}$$

$$tg\theta_3 = -\frac{1}{d} \cdot (e \cdot \sin\theta_1 + f \cdot \sin\theta_2),$$

with

$$a = 4k_1 \cdot \|A\|, \quad b = 4k_2 \cdot \|E\|, \quad d = 4k_3 \cdot \|Q\|, \quad g = 4k_4 \cdot \|B + D\|,$$
$$e = 4k_5 \cdot \|C + G\|, \quad f = 4k_6 \cdot \|F + H\|.$$

For 3—D heterogeneous media case, we can similarly analyze the stability condition via the "freezing coefficients" method [14].

SUMMARY AND CONCLUSION

The finite—difference method for modelling seismic propagation is an important tool and its stability criterion is a key for improving the computational speed of the finite—difference algorithm in 2—D or 3—D anisotropic media. However, it is lacking systematically to study the finite—difference scheme and its stability criteria for general 2—dimensional and 3—dimensional anisotropic media. In the paper, we present a fast and small storage finite—difference scheme for modelling seismic propagation in inhomogeneous anisotropic media, and systematically analyse and deduce a stability criterion for the finite—difference scheme in general homogeneous and inhomogeneous media. We believe that the results given in this paper will be very helpful in numerical simulations for anisotropic media, and provide strong theoretical evidences for the wide application of the finite difference method.

RERERENCES

1. J. Aboudi. Numerical simulation of seismic sources, *Geophysics.* **36**, 810–821 (1971).
2. A. Bayliss, K.E. Jordan, B.J. LeMesurier and E. Turkel. A fourth—order accurate finite—difference scheme for the computation of elastic waves, *Bull. Seis. Soc. Am.* **76**, 1115–1132 (1986).
3. D.C. Booth and S. Crampin. The anisotropic reflectivity technique: theory, *Gephys. J. R. Astr.*

Soc. **72**, 755–756 (1983a).

4. D.C. Booth and S. Crampin. The anisotropic reflectivity technique: anomalous arrivels from an anisotropic upper mantle, *Geophys. J. R. Astr. Soc.* **72**, 767–782 (1983b).

5. V. Cerveny. Seismic rays and ray intensities in inhomogeneous anisotropic media, *Geophys. J. Roy. Astr. Soc.*, **29**, 1–13 (1972).

6. V. Cerveny and P. Firbas. Numerical modelling and inversion of travel–times of seismic body waves in inhomogeneous anisotropic media: *Geophys. J. Roy. Astr. Soc.* **76**, 41–51 (1984).

7. K.H. Chen. Propagating numerical model of elastic wave in anisotropic inhomogeneous media: finite element methods. *The 54th SEG Annual Meeting Expanded Abstracts*, Houston, U.S.A., 631–632 (1984).

8. E.L. Faria and P.L. Stoffa. Finite–difference modelling in transversely isotropic media, *Geophysics* **59**, 282–289 (1994).

9. H. Igel, P. Mora and B. Riollet. Anisotropic wave propagation through finite–difference grids, *Geophysics* **60**, 1203–1216 (1995).

10. K.R. kelly, R.W. Ward, S. Treitel and R.M. Alford. Synthetic seismograms: A finite–difference approach, *Geophysics* **41**, 2–27 (1976).

11. B.L.N. Kennett. Theoretical reflection seismograms for an elastic medium, *Geophys. Prosp.* **27**, 301–321 (1979).

12. B.L.N. Kennett. *Seismic wave propagation in stratified media*, Cambridge University Press, Cambridge (1983).

13. D. Kosloff and E. Baysal. Forward modelling by a Fourier method, *Geophysics* **47**, 1402–1412 (1989).

14. J.F. Lu, L.Z. Gu and Z.L. Chen. *Finite–difference methods for partial differential equations*, High Education Press (1988) (in Chinese).

15. R.D. Richtmyer and K. W. Morton. *Difference methods for initial value problems*. New York, Interscience (1967).

16. C. Tsingas, A. Vafidis and E.R. Kanasewich. Elastic wave propagation in transversely isotropic media using finite differences, *Geophys. Prosp.* **38**, 933–949 (1990).

17. J. Virieux. P–SV wave propagation in heterogeneous media: Velocity–stress finite–difference method, *Geophysics* **51**, 889–901 (1986).

18. Z.J. Zhang, Q.D. He, J.W. Teng, et al.. Simulation of 3–component seismic records in a 2–dimensional transversely isotropic media with finite–difference, *Can. J. Expl. Geophys.* **29**, 51–58 (1993).

APPENDIX. LEMMAS

Lemma 1. For the equation with real coefficents
$$\lambda^2 - 2d\lambda + 1 = 0,$$
$|d| \leqslant 1$ is the sufficient and neccessary criteria of its root λ satisfying $|\lambda| \leqslant 1$.

For the proof of lemma 1, see the reference [14].

Lemma 2. If $A \in R^{n \times n}$ and $A = A'$, I is an identity matrix, then for the following equation:
$$|\lambda^2 I - 2A\lambda + I| = 0,$$
$\| A \| < 1$ is the sufficient and necessary conditions of its any constant root λ satisfying $|\lambda| < 1$.

Proof. Sufficiency:

Because A is a real symmetric matrix, there exist T^{-1}, $T \in R^{n \times n}$ to make

$$T^{-1}AT = \text{diag}(d_1, d_2, \cdots, d_n).$$

According to the equation presented in lemma 2 and above equality, we can obtain

$$(\lambda^2 - 2d_1\lambda + 1)\cdots(\lambda^2 - 2d_n\lambda + 1) = 0.$$

Because of $\|A\| \leqslant 1$ and $\rho(A) \leqslant \|A\|$, we have $\rho(A) \leqslant 1$.

According to $\rho(A) = \max_i |\lambda_i(A)|$, we have $\max_i |\lambda_i| \leqslant 1$. So, we obtain $|\lambda_i| \leqslant 1$ for any

root λ satisfying the equation presented in lemma 2.

Necessary condition: Because of $|\lambda| \leqslant 1$, and the real symmetric maxtrix $A = A'$, in terms of lemma 1, we can obtain

$$|d_i| \leqslant 1, \; i = 1, 2, \cdots, n. \; \text{i.e.} \; \rho(A) \leqslant 1.$$

Again, because A is a regular matrix, $\rho(A) = \|A\|$, i.e. $\|A\| \leqslant 1$.

Lemma 3. If $A \in R^{n \times n}$ and $A = A'$, then

(i) if $\|I - A\| \leqslant 1$, then $\|A\| \leqslant 2$;

(ii) if $A \geqslant 0$, then $\|A\| \leqslant 2$ is sufficient and necessary conditions of $\|I - A\| \leqslant 1$.

Proof. (i) Because A is a real symmetric matrix, it exists the matrices T^{-1}, $T \in R^{n \times n}$

to make $T^{-1}AT = \text{diag}(d_1, d_2, \cdots, d_n) \equiv D$

In terms of $\|I - A\| \leqslant 1$, we have $\|T(I - D)T^{-1}\| \leqslant 1$.

Obviously, I−D is a regular matrix, so,

$$\|T(I - D)T^{-1}\| = \|I - D\| \leqslant 1,$$
$$\text{i.e.} \; \max_i |1 - d_i| \leqslant 1, \; i = 1, 2, \cdots, n.$$

So, $0 \leqslant \max_i d_i \leqslant 2$, i.e. $\|D\| \leqslant 2$.

Again, because A is a regular matrix, so

$$\| A \| = \| TDT^{-1} \| = \| D \|, \quad \text{i.e.} \; \|A\| \leqslant 2.$$

(ii) The necessary condition has been proved in (i), next, we prove the sufficiency of the lemma.

According to the process of the proof above (i), we know $\| A \| = \| D \|$

Because of $\|A\| \leqslant 2$ and $A \geqslant 0$, we have $0 \leqslant \|D\| \leqslant 2$. i.e. $\max_i |d_i| \leqslant 2$.

So, we can obtain $\max_i |1 - d_i| \leqslant 1$.

$$\text{i.e.} \; \| I{-}D \| = \|T^{-1}(I - A)T\| \leqslant 1.$$

Because of the symmetry of A, We can obtain $\|I - A\| \leqslant 1$.

Proc. 30ᵗʰ Int'l. Geol. Congr., Vol. 4, pp. 65-76
Hong Dawei (Ed)
© VSP 1997

Geothermal Structure Along the Profile from Manzhouli to Suifenhe , China

JIN XU
Changchun University of Earth Science , Jilin,China,130026
EHARA SACHIO
Faculty of Engineering, Kyushu University,Fukuoka,Japan
XU HUIPING
Changchun University of Earth Science , Jilin,China,130026

Abstract

Heat flow values vary from west to east along the profile as follows: $30mWm^{-2}$ in the Wunugetu Mountain to the southwest of Manzhouli which belongs to Erguna Terrain; $59mWm^{-2}$, the average of 18 values adopted in the Hailaer Basin ;$40mWm^{-2}$, the average of 3 values measured in Duobao Mountain belonging to Da Hinggan Terrain; $70mWm^{-2}$, the average of 9 values adopted in the Songliao Basin; $47mWm^{-2}$, got from Fujin belonging to Jiamusi Terrain; and $54mWm^{-2}$, the average of 3 values measured in the Jixi Basin.The heat flow values are high in basins but low in mountains and old terrains (Erguna and Jiamusi Terrain). The correspondance among the Moho depth, high conductivity layers within the crust and upper mantle and heat flow can be found from the profile. Using Cermak and Rybach's theory, the authors calculated the transformation of P wave velocity to the heat generation density, based on the refraction data.The calculated results can be used to estimate heat flow components of the mantle and crust on the terrains. These results also show the differences in mantle heat flow among the terrains in the transect. The values vary from west to east as follwos: $23mWm^{-2}$ in the Erguna Terrain, $33 mWm^{-2}$ in the Hailaer Bansin, $33mWm^{-2}$ in the Da Hinggan mountains, $50 mWm^{-2}$ in the Songliao Basin , $25mWm^{-2}$ in the Jiamusi Terrain. It is a fact that the mantle heat flow values are very low in old Erguna and Jiamusi Terrains, very high in the Songliao Basin, and middle in the Hailaer Basin and the Da Hinggan Mountains. We may draw the conclusion that the differences in geothermal structure of the crust and upper mantle cause the change in heat flow distribution among the terrains in the transect. According to the calculated results, it is a fact that mantle heat flow and the stratum 10km above or below in the crust in every terrain have different contributions to the surface heat flow.

Keywords : heat flow , mantle heat flow, thermal structure,terrain, tectonics.

INTRODUCTION

The heat flow profile obtained in this study is nearly parrallel to the GGT profile from Manzhouli to Suifenhe , northeastern part of China . The heat flow data are useful in determining the thermal structures, thermal evolution and deep dynamic and can be used to interpret the structure of the crust and upper mantle.

Heat flow is the vertical heat flux flowing through the crust per unit area and unit time. Heat flow values are closely related to deep thermal processes, deep dynamics processes and the crust and upper mantle structure. The study of heat flow distribution clarifies not only thermal structure and processes in the crust and upper mantle , but also origin of geologic structures, especially the basin structure . This study also reveals a constraint for deriving the paleogeotherm in the basin area with regard to oil and gas development, and provides important data useful for geothermal energy development.

TEMPERATURE MEASUREMENTS IN WELLS

Wunugetu Mountain Area Near Manzhouli

We measured well-bore temperatures at two drilling sites for metal diposit exploration (Well No. 6501 and 6502) several months after cessation of drilling and obtained nearly equilibrium temperatures. The maximum depth of temperature measurement at well No. 6501 is 120 m and a linear gradient was obtained from a depth interval of 70 - 120 m . The geothermal gradient obtained is $1.00 \times 10^{-2} K/m$ .The maximum depth of temperature measurement at well No. 6502 is 70 m and a linear gradient was obtained from a depth interval of 60 - 70 m . The geothermal gradient obtained is also $1.00 \times 10^{-2} K/m$ (Fig.1). The distance between the two wells is about 50 m. The linear gradients obtained at the two sites are nearly the same , because the difference in the well head elevation is only 10 m.

Duobao Mountain Copper Mineral Zone in Nenjiang Area of Heilongjiang Province

Temperature measurements were made at six exploration wells. Equilibrium temperatures were obtained at these wells , because a very long time (about 10 years) has passed since the cessation of drilling[5]. Consequently, these wells are ideal for heat flow determination . The ID numbers of the wells are K757, ZK709, ZK819, ZK716, ZK856 and ZK842. The maximum depths at which temperatures were measured were 90 m, 235 m,70 m, 80 m, 370 m and 400 m at the above-mentioned wells, respectively. The depths at which linear temperature gradients begin are about 30 m to 40 m in all wells . The geothermal gradients obtained vary from $1.10 \times 10^{-2} K/m$ to $1.40 \times 10^{-2} K/m$. The temperature-depth relations in wells ZK709, ZK856 and ZK842 are shown in Fig.1.

Jixi Bsain of Heilongjiang Province

There are several exploration drilling wells in this area every year. The wells numbered 90-water13, water-2 and 93-156 were chosen for temperature gradient measurement . The well 90-water13 as drilled for several years, and water-2 for 3 months, temperatures therefore in these bore-holes were regarded to be equilibrium. The 93-156 finished drill for only fifty-five hours. The maximum temperature measuring depths in 90-water13, water-2 and 93-156 are 410m , 390m and 500m,respectively, the temperature gradients from 170m to 370m, 150m to 390m and 120m to 480m are $3.7 \times 10^{-2} K/m$, $3.9 \times 10^{-2} K/m$ and $2,9 \times 10^{-2} K/m$,respectively. The temperature-depth relations in wells 90-water13, water-2 and 93-156 are shown in Fig.1.

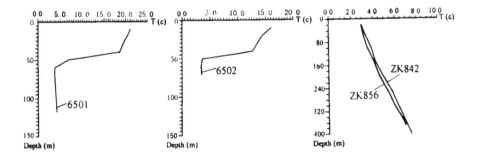

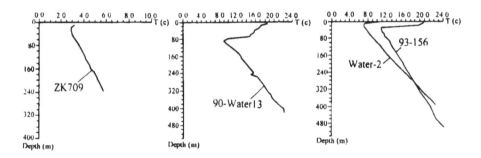

Fig.1 The temperature-depth relation at wells 6501 , 6502, ZK842,ZK856,ZK709,90-water13, water-2 and 93-156.

MEASUREMENT OF THERMAL CONDUCTIVITY

Four core samples were selected from Manzhouli 6501 bore-hole. Eight,five , five, three, four, five and ten core samples were selected from bore-holes ZK842, ZK856 , Zk525, 93-15, 88-water-4, water-2 and 93-156, respectively. Thermal conductivity of these rocks were measured by using a ring source probe thermal conductivity apparatus in the Geothermal Laboratory of the Institute of Geology , Academia Sinica. The results are summarized in Table 1.

HEAT FLOW DETERMINATION

Heat flow is calculated by the following equation :
$$q = - k \, (dT/dz) \tag{3}$$
where q is the heat flow (unit : mW/m^2), k is the thermal conductivity (unit :W/mK) and dT/dz is the geothermal gradient (unit: K /m).Linear geothermal gradients are estimated at 1.0×10^{-2}K/m, 1.3×10^{-2}K/m, 1.4 K/m ,1.1×10^{-2}K/m, 3.7×10^{-2}K/m 4.3×10^{-2}K/m, and 2.9×10^{-2}K/m,for wells 6501 , ZK842 ,ZK856 ,ZK709,90-water-13,water-2 and 93-156 ,respectively. By using the thermal conductivity obtained above, heat flow values are calculated based on Eq. (3) . The resulting heat flow values are estimated as follows: 30 mW/m^2 for well 6501, 40mW/m^2 for ZK842 , 41mW/m^2 for ZK856 , 40mW/m^2 for ZK709, 57mW/m^2 for 90-water-13, 70mW/m^2 for water-2

and 35mW/m^2 for 93-156. The thermal conductivity used for ZK709 were taken from a site ZK525 which is located 200m away from it.

Previously, eighteen heat flow values were reported from the Hailaer Basin[5] and ten values for the Songliao Basin[12,13] along the GGT profile. The heat flow values for all the data mentioned above are listed in Table 2. The distribution of heat flow is shown in **Fig.2**.

Table 1 The results of thermal conductivity measurements

NO	Depth (m)	Conduc. (W/mK)	Rock
6501			
	87	3.4123	granite
	100	3 4740	granite
	105	2 7795	granite
	114	2 3307	granite
ZK842			
	100	3 1629	tuff lava
	199	3 2495	tuff lava
	200	3 5010	tuff lava
	300	2 1991	tuff
	399	3 4719	andesite
	400	3 4324	andesite
	500	2 2557	andesite
	613	2 6025	andesite
ZK856			
	100	3 6142	tuff lava
	200	2 9813	tuff lava
	236	3 2840	tuff lava
	236	3 0344	tuff lava
	·350	2 3266	tuff lava
ZK525			
	82	4 3762	tuff lava
	118	3 5238	tuff lava
	167	3 6280	tuff lava
	216	3 2158	tuff lava
	233	3 3747	tuff lava
93-156	56	1 469	basalt
	108	0.504	tuff
	150	1 237	tuff
	220	1 452	sandrock
	250	1 132	sandrock
	309	1 164	tuff
	333	1 158	tuff breccia
	380	1 188	tuff breccia
	439	1 147	tuff breccia
	500	1 547	sandrock
water-2		1 411	sandrock
		1 857	tuff
		1 620	sandrock
		1 568	tuff breccia
		1 675	sandrock
93-15			
	75	1 681	basalt
	179	1 657	basalt
	370	2 020	sandrock
88-water-4		1 462	basalt
		1 332	basalt
		1 417	sandrock
		0.764	sandrock

Table 2. The heat flow data along and near the GGT profile

No	Loca	E.lo	N lat	Collar elev. (m)	Depth range (m)	Gradient G ±SD (K/Km)	Conduc K±SD(n) (w/m k)	Heat flow (mW/m²) Obs. Corr.
1	WUNUGETU MOUNTAIN	117°14'	49°26'	846 5	70 -120	10.0	3 00	30 0
2	HAILAER WUBEI(1)	117°37'	48°40'	576 5		40 6	1 66	67.4
3	HAILAER WUBEI(2)	117°46'	48°34'	573.3		40.6	1 72	69.9
4	HAILAER WUBEI(3)	117°43'	48°32'	576 2		41 1	1 69	69 5
5	HAILAER WUBEI(4)	117°55'	48°35'	562 5		45 4	1 57	71 2
6	HAILAER WUBEI(5)	117°50'	48°33'	571 2		37 8	1 65	62.4
7	HAILAER WUZONG	117°42'	48°26'	576.1		39 1	1 62	63.2
8	HAILAER WUNAN(1)	117°53'	48°10'	588.1		34.2	1.68	57 4
9	HAILAER WUNAN(2)	117°47'	48°18'	588.0		39.7	1 71	67.8
10	WUNAN(3)	117°47'	48°13'	590 9		31.4	1.84	57.9
11	WUNAN(4)	117°43'	48°13'	597.3		33.7	1 60	54.0
12	WUNAN(5)	117°41'	48°10'	599 2		33 2	1 73	57 4
13	HAILAER HONGQI	118°01'	48°53'	558 2		34 2	1 63	55 7
14	CAGANNUOER	116°28'	47°55'	562 1		33.9	1 65	56 1
15	HAILAER BEIE(1)	117°20'	48°06'	574 9		31 3	1 85	57 9
16	HAILAER BEIE(2)	117°09'	47°51'	614 5		30 0	1 95	58 6
17	HAILAER BEIE(3)	117°06'	47°59'	587 2		33 6	1 62	54 4
18	HAILAER FUHE	118°51'	48°10'	742 8		36 6	1 65	60.3
19	WUGUNUOE	118°55'	49°00'	659 2		40 1	1.58	63.2
20	TONG MOUNTAIN(1)	125°49'	50°14'	529 8	20 - 370	14.0	2.91(5)	40.7
21	TONG MOUNTAIN(2)	125°50'	50°13'	510 3	20 - 400	13.0	3.09(8)	40.1
22	DUOBAO MOUNTAIN	125°48'	50°14'	482 8	20 - 235	11.0	3 62(5)	39.8
23	TAIKANG	124°23'	46°38'		1200 - 1540	44 5	1 56(7)	69.5
24	DAQING	124°34'	46°09'		1480 - 1700	56 8	1 67(5)	95.0
25	ANDA	124°43'	46°01'		2480 - 3900	32 4	2 50(5)	80 8
26	DAQING	124°51'	46°27'		1400 - 3025	36 4	2.31(4)	84 2
27	ZHAOZHOU	124°58'	45°54'		1050 - 3175	33 7	2 23(6)	75 4
28	MINGSHUI	125°43'	47°03'		760 - 860	41 3	1 36(3)	56 1
29	KESHANG	125°44'	48°01'		500 – 800	33 3	1 33(3)	44.4
30	QINGGANG	125°49'	45°46'		640 – 740	40 0	1 13(3)	45 2
31	HULAN	126°21'	45°58'		400 – 800	47 8	1 62(2)	80 4
32	FUJIN	132°00'	47°00'		525 - 2500	20 5	2 27(5)	46 5
33	JIXI(1)	131°09'	45°08'		80 - 410	37 0	1 53(5)	56 8
34	JIXI(2)	130°59'	45°07'		30 - 500	29 0	1 21(10)	35 0
35	JIXI(3)	130°53'	45°05'		40 - 390	43 0	1 62(5)	69 9

CALCULATION OF THE VERTICAL DISTRIBUTION OF HEAT PRODUCTION IN THE CRUST

In order to set up heat structure model on the terrains from geothermal data , the authors studied the theory on statistical relations between seismic wave velocity and the heat production, which was advanced by Cermak[2,3,4] and Rybach[9]. According to the theory the authors made the calculation of the transformation from P wave velocity to the heat production ,based on the refraction data[7]. The calculated results can be used to estimate heat flow components of the mantle and crust on the terrains.

In five different areas, the Vp-A relationships were used for conversion calculations. They are Wunugetu Mountain area near Manzhouli, Hailaer Basin, Duobao Mountain area, Anda and Jixi Basin (Fig.3,4,5,6,7).

70

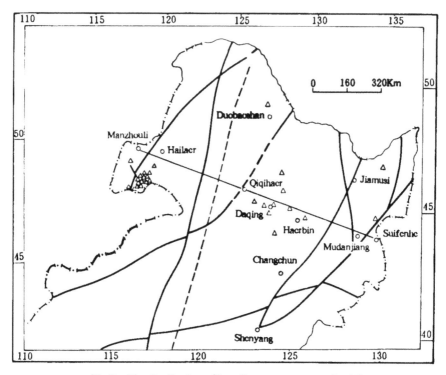

Fig.2 The distribution of heat flow measurement sites(Δ).

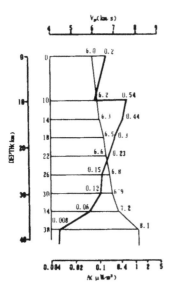

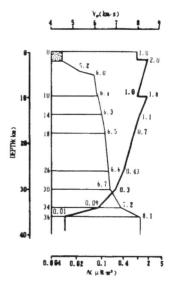

Fig.3 Seismic velocity-heat production conversion in Wunugetu Mountain area.

Fig.4 Seismic velocity-heat production conversion in Hailaer Basin.

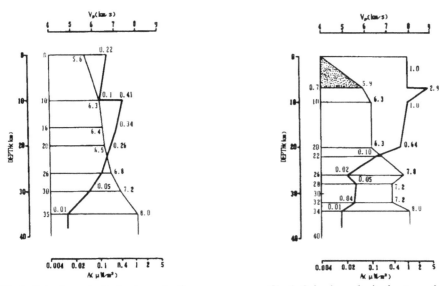

Fig.5 Seismic velocity-heat production conversion in Duobao Mountain area.

Fig.6 Seismic velocity-heat production conversion in Anda Basin.

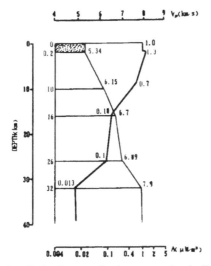

Fig.7 Seismic velocity-heat production conversion in Jixi Basin.

THE THERMAL STRUCTURE OF THE CRUST AND UPPER MANTLE

One-Dimensional Thermal Structure In steady-state Conditions
Using the one-dimensional thermal conduction equation in steady-state condition, the temperatures and heat flows can be obtained as below (Table 3).

Two-Dimensional Thermal Structure In Steady-state Condition
It is known to us, 2-D finite element simulation is a powerful method in the study of thermal structure of the crust and upper mantle and is also suitable for the study of two-dimensional temperature distributions in GGT profile. The difference between

them is the different parameters which are adopted in the calculation .Because of the uncertainty of parameters of the whole Europe-Asia and Pacific plate and some terrains in the profile, such as the plate converging velocity ,the fault movement speed, the erosion rate; the shortening ,the increasing ,the stretching of the plate etc. the 2-D finite element modeling is constrained only by steady-state conditions.

Table 3　Mantle heat flow and other parameters

Loca.	Surface heat flow (mW/m^2)	Moho depth (km)	Heat flow upper 10km (mW/m^2)	Contribution of lower crust (mW/m^2)	Mantle heat flow (mW/m^2)	Moho temperature (°C)	Average mantle heat flow (mW/m^2)
Erguna	30.0	38	1.3	6.1	22 6	319	22.6
Hailaer Basin	71.2	38	17.9	16 6	36.7	819	
	55.7	36	14.3	15.6	25.9	624	32.5
	60.3	37	15.2	14.0	31 1	726	
	63.2	36	16.0	11.0	36.2	785	
Da Hinggan							
Mountain	40.1	35	1.6	5.3	33.2	402	33.2
Songliao	69.5	34	23.2	11 1	35.1	621	
Basin	95 0	35	31.1	6.7	57.2	861	
	80.8	34	15.4	9.9	55.5	795	49.8
	84.2	35	26.9	6.0	51.3	811	
	75.4	34	16.9	10.5	48.0	736	
	80.4	32	19.7	8.8	51 9	828	
Jiamusi	56.7	32	22.1	4.2	30.4	602	
Terrain	35.0	32	9.7	5.3	20.0	475	25.2

The purpose of the 2-D finite element modeling is mainly to clarify the two-dimensional thermal structure of the crust and upper mantle. Thus the simplified model is derived from the structure model obtained from other geological and geophysical methods. In this paper, the simplified model is based on the consideration of the reflection and refraction seismic interfaces, sedimentary, crystalline bedrock and Moho as the base surfaces.　The terrains (Erguna , Da Hinggan , Jiamusi and Xinkai Terrain) and basins(Hailaer and Songliao Basin) are also taken into consideration.

The 2-D finite element equation is:
$$\rho c(\partial T/ \partial t) = k(\partial^2 T/\partial x^2 + \partial^2 T/\partial z^2) \quad + \quad A \qquad (4)$$
where ρ is the material density , c is specific heat, k is thermal conductivity, and A is radiogenic heat production.In steady-state conditions, that is to say, temperature varies independent on time, the equation is:
$$k(\partial^2 T/\partial x^2 + \partial^2 T/\partial z^2) \quad = \quad -A \qquad (5)$$
and it can also be represented by an minimized functional formula as follow:
$$I = 1/2\int_s k(\partial^2 T/\partial x^2 + \partial^2 T/\partial z^2)ds + \int_s ATds + \int_\tau q_n dl \qquad (6)$$
where I is energy functional, s is the study region , τ is the boundary of the region , q_n is the heat flow through the boundary, k is the thermal conductivity and A is radiogenic heat production.　According to these equations, if k, A and the boundary condition are known, the distribution of the temperatures can be obtained. The result is showed in Fig.8.

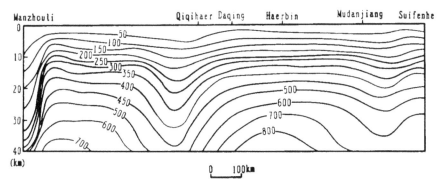

Fig.8 Vertical temperature distribution from Manzhouli to Suifenhe.

CONCLUSIONS AND DISCUSSIONS

Heat flow values vary from west to east along the profile as follows: 30mWm^{-2}, in the Wunugetu mountain to the south-west of Manzhouli which belongs to Erguna Terrain; 59 mWm^{-2}, the average of 18 values adopted in the Hailaer Basin; 40mWm^{-2}, the average of 3 values measured in the Duobao mountain belong to Da Hinggan Terrain; 70mWm^{-2}, the average of 9 values adopted in the Songliao Basin; 47mWm^{-2}, got from Fujin belonging to Jiamusi Terrain, and 54 mWm^{-2}, the average of 3 values measured in the Jixi Basin . The changes in heat flow along the profile are shown in Fig.9. It is a common knowledge that heat flow measurement is limited by the distribution of drillings being used for temperature logging. There is a shortage of measured heat flow values in Da Hinggan and Zhangguangcai Mountains near the transect. So the changes of heat flow in the section are shown by dotted line in the Fig.9. The heat flow values in basins are high but low in mountains and old terrains(Erguna and Jiamusi Terrain).

There is a close relation between Moho depth and heat flow . It is a general rule that the higher the heat flow value, the smaller the Moho depth is[1]. The Moho surface ,as located by refraction seismic survey in the transect[7], is about 40 km deep in west Erguna Terrain, 37km in the Hailaer Basin, 33 km in the Songliao Basin, and 39 km in east Jiamusi Terrain. The scatter of the heat flow values above-mentioned corresponds to the Moho depth directly (Fig.9). The correspondence can be found easily in other areas in China[6].

The relation between heat flow values and the position of high conductivity layers within the crust(CHCL) and upper mantle (UMHCL) is close yet. This relation is presented in Adam's experienced equation: $h=h_0 q^{-a}$, where h is depth of high conductivity layer, q is tectonic heat flow value, h_0 and a superscript are parameters which represent structure characters of an area. Fig.9 also shows the different depths to high conductivity layers whithin the crust and the upper mantle, which are derived from the magnetotelluric sounding in the transect[7]. The rising and falling of heat flow in the Da Hinggan and Zhangguangcai Mountains, where heat flow values have not been measured, can be outlined based on changes of the Moho depth and high conductivity layers in the crust and upper mantle.

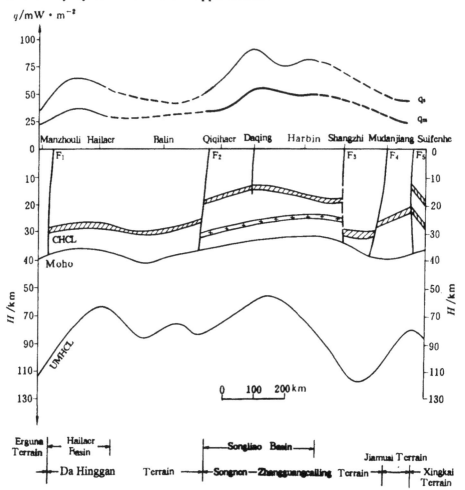

Fig.9. Plots of changes of surface heat flow , Mantle heat flow, and geology structure along the transect. F1: Deerbugan Fault; F2: Nenjiang Fault ;F3: Jiayi Fault; F4: Mudanjiang Fault;F5: DunMi Fault; CHCL: crust high conductivity layer; UMHCL: upper mantle high conductivity layer.

There are obvious differences in the heat flow distribution between the Songliao Basin and the Hailaer Basin. The values in the Songliao Basin are higher than those in the Hailaer Basin, and exceed the world average value (63 mWm^{-2})[8].That means a great difference in forming mechanisms and structure characters between the two basins. The existence of high heat flow values in the Songliao Basin is caused by three factors besides shallow Moho and the upper mantle[13]. Firstly, the Caledonian,Variscan and Austrian granites are distributed in the Songliao Basin having high heat generation . Secondly, the Songliao Basin in East China is a part of rift valley series around the Pacific, and is a compound basin of rift and depression. The extension of the crust makes the heat movement active; Finally, the Songliao

Basin is a sealed stagnant basin. There is no sluiceway in the area. The flowage of groundwater is slow and the flow velocity is only 6.1 mma^{-1}. Thus the heat can not escape easily. It is important that the large area covered by argillite formed by rapid lake immersion in the basin has better natures of gathering and isolating heat. So the heat coming from deep earth could be conserved for a long time.

These results also show the differences in mantle heat flow among the terrains in the transect. The values vary from west to east as follows: $23mWm^{-2}$ in the Erguna Terrain, $33mWm^{-2}$ in the Hailaer Basin, $33mWm^{-2}$ in the Da Hinggan Mountains, $50mWm^{-2}$ in the Songliao Basin, and $25mWm^{-2}$ in the Jiamusi Terrain. It is a fact that mantle heat flow values are very low in old Erguna and Jiamusi terrains, very high in the Songliao Basin, and intermediate in the Hailaer Basin and the Da Hinggan Mountains. We come to the conclusion that differences on geothermal structure of the crust and the upper mantle cause the change in heat flow distribution among the terrains in the transect. According to the calculated results, we know the fact that mantle heat flow and the 10km above or below in the crust in every terrain have different contributions to the surface heat flow.

There are two large basins in the transect -- the Songliao and the Hailaer Basins. However, there exist great differences not only in mantle heat flow, but also in geothermal structure of the crust between the basins. They are not the same type. Mantle heat flow value is $50mWm^{-2}$ in the Songliao Basin, which is 60 percent as much as the surface heat flow value. It is $33\ mWm^{-2}$ in the Hailaer Basin, which is only 52 percent as much as the surface heat flow value. The great change in the values shows a great difference of the upper mantle geothermal structure in the basins. Their geothermal structure varies with the tectonic locations of basins. A Japanese scholar presented his study result after comparing the mantle structure with the history of plate underthrusting. The study is based on the theory of seismic tomography[10]. Following his view we obtain the interpretation of forming high mantle heat flow values in the Songliao Basin: a piece of the Pacific retentive cold plate subducts into the mantle below Songliao Basin. When the piece falls from the upper mantle to the lower mantle, heat plume produced in the mantle will rise upward to compensate for the fall. The rising movement of "hot" plume probably has formed high mantle heat flow in the Songliao Basin. Calculated geothermal structure within the crust shows that the upper crust of the uppermost above 10 km contributes much more to the surface heat flow($22.2\ mWm^{-2}$) than that in the lower crust of the lowermost 10 km ($8.8mWm^{-2}$) in the Songliao Basin. On the other hand, the contribution of upper and lower crust are roughly equal in the Hailaer Basin (they are $15.8\ mWm^{-2}$ and $14.3\ mWm^{-2}$, respectively). The fact reveals that the enrichment of heat generative radioelements toward the surface is larger in the Songliao Basin, but the enrichment is less in the Hailaer. Paying attention to lithologic characters, we find that the difference degree among acid, basic and ultrabasic rocks in the Songliao Basin is higher than that in the Hailaer Basin. The above-mentioned situations are consistent with the fact that huge granitic intrusive probably exist in the depth range from 5 to 13 Km beneath the Songliao Basin[13].

Acknowledgements

We thank Wang Jiyang and Shen Xianjie (Institute of Geology, Chinese Academy of Sciences) for their help , also thank Shi Yaolin (Graduate School , Chinese Academy of Sciences) for his help in 2-D finite element calculation, and Zhou Yunxuan for his review of our manuscript.

REFERENCES

1. D. S. Chapman and L. Rybach.. Heat flow anomalies and their interpretations, *J.Geodynamics* **4** , 3-37(1985).
2. V. Cermak. Crust heat production and mantle heat flow in central and eastern Europe, *Tectonophysic* **159:**314,195-215(1989).
3. V. Cermak. Crustal temperature and mantle heat flow in Europe, *Tectonophysic* **83**, 123-142 (1989).
4. V. Cermak and L. Rybach. Vertical distribution of heat production in the continental crust, *Tectonophysic* **159**,217-230 (1989)
5. Exploration Department of Daqing Petroleum Administrative Bureau. The report in study of oil rocks in Hailaer Basin, Neimenggu (in Chinese) (1990).
6. Jin Xu and Gao Rui. Thermal and tectonic structures of the crust and upper mantle in eastern China (in Japanese),*Journal of the Geothermal Research Society of Japan* **12:2**, 129-143(1990).
7. Jin Xu, Yang Baojun. The study of characters on geophysical field and deep geostructure along the geoscience transect from Manzhouli to Suifenhe, China, Seismology Press, Beijing(1994).
8. W.H.K. Lee. On the global variations or terrestrial heat flow, *Phys. Earth. Planet.* Inter. **2**, 332-359(1970).
9. L. Rybach , G. Buntebarth. The variation of heat generation density and seismic velocity with rock type in the continental lithosphere, *Tectonophysics* **103**,335-344(1984).
10. Shigenori Maruyama etc. Plume tectonics, *Science* **63:6**,373-386. (in Japanese)(1993).
11. The Second Regional Geological Survey of Heilongjiang province. The report of detail survey on Duobao mountain cupper mineral zone in Nenjiang area of Heilongjiang province (in Chinese) (1982).
12. Wang Jiyang and Huang Shaopeng. Compilation of heat flow data in the China continental area (2nd Edition), *Seismology and Geology* **12 : 4** , 351-366(1990).
13. Wu Qianfan and Xie Yizhen . The heat flow of SongLiao basin(in Chinese), *Seismology and Geology* **7:2** , 59 - 63(1985).

Proc 30th Int'l. Geol. Congr., Vol. 4, pp. 77-90
Hong Dawei (Ed)
© VSP 1997

Global Geoscience Transect Programme in Finland: The GGT/ SVEKA Transect

Kalevi Korsman[1], Toivo Korja[2] and Petri Virransalo[1]

[1] *Geological Survey of Finland, FIN-02150, Espoo, Finland*
[2] *University of Uppsala, Department of Geophysics, Uppsala, Sweden*

Abstract

The studies along the SVEKA profile form a Finnish contribution to the Global Geoscience Transect Programme. The GGT/SVEKA transect crosses the boundary zone between the Archaean continent and the Palaeoproterozoic Svecofennian Island Arc Complex and the main units of the Svecofennian Island Arc Complex. The major focus in the GGT/SVEKA program has been to study the evolution and structure of the lithosphere based on geophysical and geological observations with special emphasis on the relationship between crustal thickness variations and a strong Palaeoproterozoic heat flow as well as on the geological significance of deep crustal conductors. The crustal thickness variations and high heat flow are attributed to magmatic underplating. The geometry and position of the deep conductors in southern Finland indicates a collision between two tectono-metamorphic belts.

1. Introduction

The studies along the SVEKA profile form a Finnish contribution to the Global Geoscience Transect Programme. The GGT/SVEKA transect, from which the most comprehensive geological, geochemical and geophysical data in Finland are available, is located in the central part of the Fennoscandian Shield covering an 160 km wide and 770 km long area from the Finnish-Russian border in northeast to the Finnish-Swedish border in southwest in the Åland archipelago (Fig 1). The GGT/SVEKA transect crosses the Archaean-Proterozoic boundary zone and the main tectonic units of the northern part of the Svecofennian Domain and is therefore well suited to investigate the Svecofennian orogeny and its effects in the Archaean Karelian crust.

The Svecofennian orogeny produced one of the thickest continental crust in the world excluding the active orogenic regions. The crust comprises of juvenile Palaeoproterozoic material, is characterized by low p - high T metamorphism and lacks

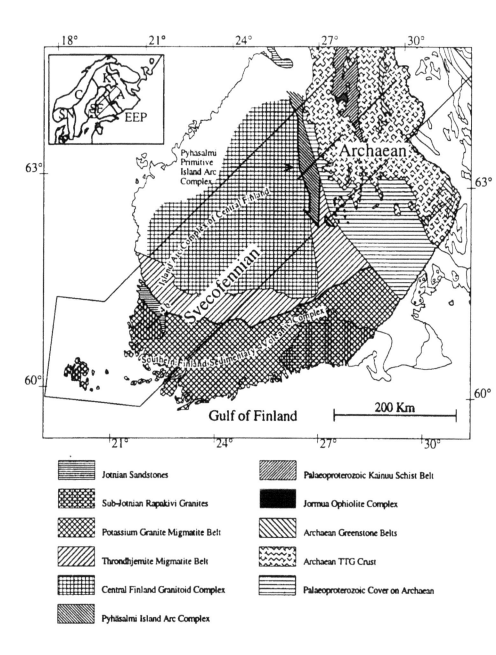

Fig. 1. Main tectonic units in Southern Finland. Lines 4a and 4b denote the location of cross-sections shown in Figures 4a and 4b, respectively. Inset: Main tectonic units of the Fennoscandian Shield and the location of the GGT/SVEKA profile. A-Archaean Domain, K-Karelian Province, SF-Svecofennian Domain, SN-Sveconorwegian Domain, C-Caledonides, EEP-East European Platform.

any exposed sections of deep crustal rocks. The crust contains also two major crustal conductors traversing cross the entire shield. The seismic structure of the Svecofennian crust is characterized by the presence of a high velocity ($v_p > 7$ km/s) layer at the bottom of the crust that accounts for most of the large and spatially rapid crustal thickness variations. The regions of thickest crust lacks topographic and gravity expressions to compensate and explain the large variations in the crustal thickness. Nor is there any evidence of major tectonomagmatic activity after the emplacement of rapakivi granites implying that the crustal structure that we image today was formed in and has been preserved since the Svecofennian orogeny and the emplacement of the rapakivi granites more than 1.5 Ga ago.

The major focus in the GGT/SVEKA program has been to study the evolution and structure of the lithosphere with a special emphasis on the crustal thickness variations, the high heat flow of the Svecofennian orogeny and the significance of deep crustal conductors. In this study we have considered the following problems: (i) How and when the crust was thickened into its present thickness in the eastern part of the transect? (ii) Why and how the crustal thickness variations were preserved until now? (iii) What is the relationship between the high heat flow of the Svecofennian orogeny and deep crustal structure? (v) What is the geological significance of the crustal conductors found in southern Finland?

2.   Major geological and geophysical features along transect

The Archaean crust on the GGT/SVEKA transect is composed mainly of Mesoarchaean migmatites and Neoarchaean tonalites and granodiorites. Greenstone belts form only narrow zones compressed between the granitoid complexes. Palaeoproterozoic Svecofennian Island Arc Complex is characterized by granodiorites and migmatites, whereas the amount of volcanic rocks is small. The evolution of the Palaeoproterozoic bedrock is especially well represented along the GGT/SVEKA transect. Major crust forming and deforming events are listed in Table 1.

The easternmost part of the GGT/SVEKA transect is located in the Archaean granitoid-greenstone area where the transect crosses the Kuhmo Greenstone Belt. Further to the southwest the transect crosses the Palaeoproterozoic Kainuu Schist Belt that hosts the allochthonous Jormua ophiolite complex. The Jormua complex is among the oldest ideal oceanic crust identified so far in the world [1]. The boundary zone between the Archaean continent and Svecofennian Domain is sharp. According to Nd-Sm data there is no Archaean crust westwards from the boundary zone [2]. The primitive Pyhäsalmi Island Arc complex lies close to the margin of the Archaean continent. Most of the Svecofennian Domain in Finland is composed of the two mature island arc systems in southern and central Finland. The sharp boundary zone of these two belts is E-W directed in southern Finland. The Mesoproterozoic Rapakivi batholiths are situated in the southwestern part of the GGT/SVEKA transect. The deposition of the Jotnian sandstone, close to rapakivis, may have started prior to the emplacement of rapakivi granites [3]. The sandstone is cut by Subjotnian diabase dykes, which represent the youngest rocks in the GGT/SVEKA transect area.

The Fennoscandian Shield is characterized by exceptionally thick crust but also by large and sometimes spatially rapid variations in crustal thickness ([4,5] and Fig 2). Thinnest crust (35-45 km) is found in the Archaean Karelian Province and in a few roughly EW directed zones. The high velocity layer is thin or nearly absent in these regions (Fig. 3) and the Moho is characterized by strong, nearly subhorizontal reflectors. The regions of the thickest crust (45-65 km), on the other hand, are characterized by thick lower crustal high velocity layer with decreasing reflectivity towards the Moho boundary where a weakly reflective Moho is encountered. The thickness of the high velocity layer reaches nearly 30 km km beneath the Archaean Karelian boundary zone in SE Finland (Fig. 3). The GGT/SVEKA transect [6] provides a seismic crustal structure that is representative for the Svecofennian crust in general. The Archaean Karelian Province east of the Kuhmo Greenstone Belt has a crustal thickness of ca. 42 km with a thin lower crustal high velocity layer. The thickness of both the crust and the lower crustal high velocity layer increases towards the Archaean-Proterozoic boundary zone reaching the maximum

Age (Ga)		Event
3.1	- 2.7	Evolution of the Archaean Greenstone Belt
2.2	- 1.97	Continental rifting of the Archaean continent
1.954		Opening of the initial ocean - Jormua ophiolite
1.93	- 1.915	Primitive island arc magmatism
1.91	- 1.885	Collision between the Archaean continent and the
		Palaeoproterozoic Svecofennian island arc complexes
1.885		Dry basic and Opx bearing granitic magmatism in the
		boundary zone - high T - low P metamorphism
1.91	- 1.888	Mature island arc volcanism
1.88	- 1.86	Accretion of the island arc complexes in southern Finland
1.888	- 1.870	High T - low P metamorphism in central Finland
1.84	- 1.812	High T - low P metamorphism in southern Finland
1.588	- 1.540	Rapakivi magmatism
		Jotnian sandstone
1.26	- 1.25	Postjotnian diabases

Table 1. Major geological events along the GGT/SVEKA transect.

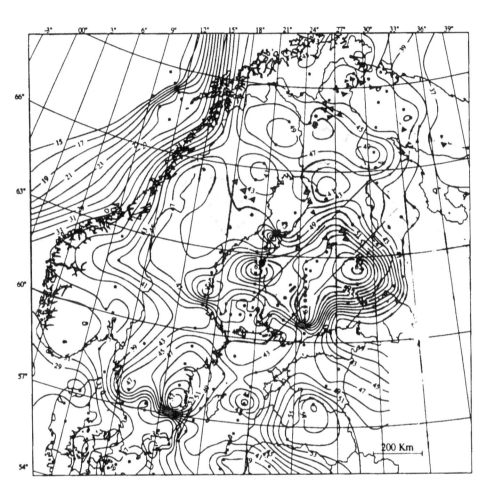

Fig. 2. Crustal thickness and crustal conductors in Fennoscandia. Contours represent depths in km from the Earth's surface to the Moho boundary obtained from refraction seismic studies. The Moho map is modified from the map of [4] by including new data from [6,9,10,11,12,13,14]. Moho depths are interpolated from 2D velocity models at sites shown as black dots. Gray areas indicate exposed parts of crustal conductors revealed by magnetovariational [15], magnetotelluric [7] and airborne electromagnetic data (GSF) from the central and northeastern parts of the shield; no data available from the southeastern part of the shield. Black triangulares show the dip direction of conductors. The site of the GGT/SVEKA transect is as a polygon.

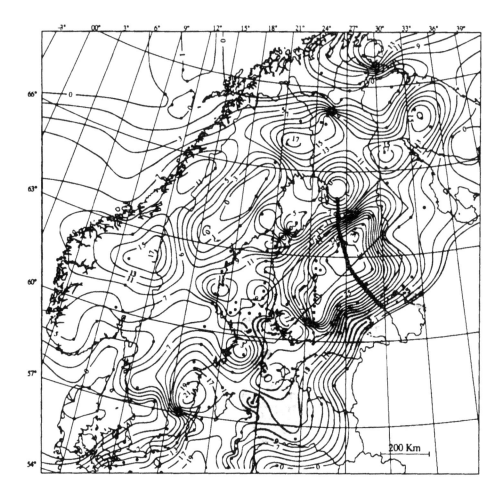

Fig. 3. Thickness of the lowermost high velocity crustal layer. Contours represent thickness in km of the layer in which seismic P-wave velocities are between 7.0 and 7.7 km/s. The map is modified from the map of [4] by including new data from [6,9,10,11,12,13,14]. Thickness values are interpolated from 2D velocity models at sites shown as black dots. Gray areas show rapakivi intrusions. The boundary between the Archaean continent and the Svecofennian Domain is hatched. The site of the GGT/SVEKA transect is as a polygon.

values of 59 km and 21 km, respectively, right beneath the suture zone. Further to the SW, as far as the region of the Subjotnian rapakivi intrusions, the crust and high velocity layer thin slowly to 53 km and 15 km. In the region of the Subjotnian rapakivi intrusions, the crust thins abruptly to ca. 45 km associated with the thinning of the lower crustal layer to 10 km or less (Fig. 2 and 3). The velocity variations are rather smooth but considerable variations in velocity ratios occur along the transect. In the upper mantle the v_P/v_S ratio is about 1.73 beneath the both ends of the transect whereas beneath the Central Finland Island Arc Complex ca. 1.78. In the lower crustal high velocity layer the velocity ratio ranges from 1.76 to 1.79, the highest values being observed beneath the Central Finland Island Arc Complex. The ratio has a nearly constant value of 1.77 in the middle crust all over the Palaeoproterozoic Svecofennian Domain and in the Archaean Karelian Province as far as to the Kuhmo Greenstone Belt in east. East of the Kuhmo Greenstone Belt the velocity ratio at corresponding depths (20-40 km) is, however, only ca 1.74.

Electrically the Fennoscandian Shield is characterized by large resistive regions with intervening long, narrow belts of crustal conductors ([7] and Fig. 2). The conductors are mostly located in the upper and middle crust with occasional penetration into the lower crust as in the Skellfeteå region in northern Sweden [8]. The airborne electromagnetic mapping in Finland indicate that most of these conductors are exposed. Due to the large conductances involved and the correlation with surface geology the conductivity mechanism invoked to explain the enhanced conductivity is primarily electronic as found in graphite- and sulphide-bearing metasedimentary rocks. Two of these conductive belts cross the GGT/SVEKA profile. The northeasternmost is associated with the Archaean-Proterozoic boundary zone and extends from Russia to the Bothnian Bay. The second crosses the transect roughly at the southern Throndhjemite Migmatite Belt and extends from the Archaean-Proterozoic boundary zone in the Finnish Karelia to the Skellefteå region in northern Sweden. In southern Finland, the conductor is narrow (50 km) and nearly vertical extending to the depths of about 20 km whereas in northern Sweden it is about 300 km wide, dips towards NE and penetrates the entire crust to the depths of about 40 km. The airborne survey has imaged several elongated near-surface conductors on the top of the western part of the deep conductor indicating that the deep conductor is exposed. In the east, the deep conductor becomes narrower and has no surface expression according to the airborne data. To the north, however, a region of near-surface conductors, similar to those in western Finland, is observed.

3. Thickening of the crust and temporal variations in the heat flow in the eastern part of the SVEKA profile

The crust of the shield reaches it maximum value of over 60 km along the collision

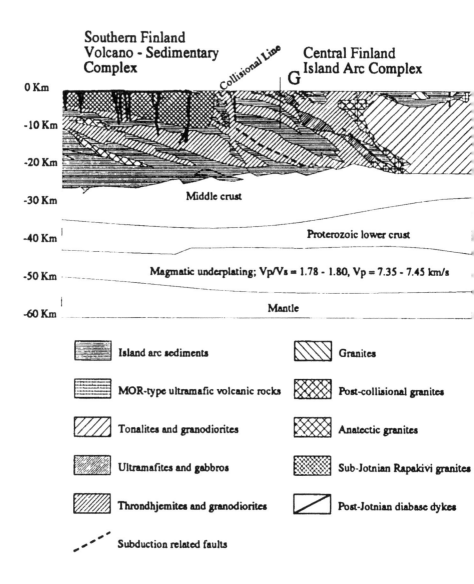

Fig. 4a. Crustal cross section based on integrated interpretation of geophysical and geological data across the boundary zone between the Central Finland Island Arc Complex (to the right) and Southern Finland Volcano-Sedimentary Complex (to the left) (see Fig. 1).

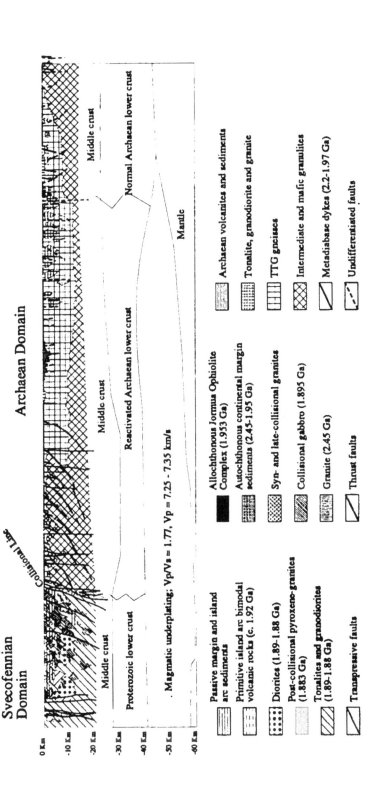

Fig. 4b. Crustal cross section based on integrated interpretation of geophysical and geological data across the boundary zone between the Archaean Domain (to the right) and the Palaeoproterozoic Svecofennian Domain (to left left) (see Fig. 1).

zone between the Svecofennian and Archean continents whereas the Archaean crustal thickness is only ca. 40 km. Based on structural observations the crust first thickened tectonically [16]. This is also corroborated by the presence of deep crustal conductors along the boundary zone that require a transportation of carbon-bearing sedimentary material at great depths. After tectonic thickening gabbros and over heated orthopyroxene bearing granites intruded 1.885 Ga ago [17,18]. According to geochemical data gabbros are melting products of mantle. Fusion of mantle and upwelling of magma caused melting of the lower crust, which produced dry Opx bearing granite intrusions also in the upper parts of the crust [17,19,20]. High T - low P-metamorphism (T = 800 C°, P = 5 kb) is connected partly with hot fluids caused by mantle melting and Opx bearing granite intrusions. Gabbros, Opx granites and high temperature events of metamorphism have the same ages, 1.885 Ga. Thus the thickening and evolution of the lower crustal high velocity layer and the high temperature-low pressure metamorphism are associated with the magmatic underplating.

In the collision zone between the Archean and Svecofennian continents, the crust is not only thick but also dense based on gravimetric data [21]. The increase in density is caused by basic intrusions situated mostly below the present erosion surface. Thick crust was preserved because of its increased density caused by the magmatic underplating after the tectonic thickening of the crust.

The thick high velocity layer continues also into the Archean side and this part of the Archean crust is overprinted by a relatively strong Svecofennian metamorphism. In contrary, in the easternmost part of the GGT/SVEKA transect, where the crust has a normal thickness of only 40 km and the high velocity lower crustal layer is nearly absent, the metamorphic overprinting of the Svecofennian orogeny is only weakly recognizable [22,23]. According the aforementioned observations also the western part of the Archean lower crust has been influenced by magmatic underplating, which is indicated in upper parts of the crust by the overprinting of the Svecofennian metamorphism

4. Accretion of the two island arc complexes in southern Finland

The collision zone between the Central and Southern Finland Island Arc Complexes is situated in the southern part of the GGT/SVEKA transect and coincides with the Southern Finland conductive belt. The Central Finland Island Arc Complex is characterized by throndhjemite migmatites and the Southern Finland Complex by potassium granite migmatites. The age of the metamorphic peak in the throndhjemite migmatite belt is 1.885 Ga and in the potassium migmatite belt between 1.885-1.810 Ga [17,24,25]. The difference of the neosome between two belts was caused by different compositions of migmazised sediments. Sediments of the throndhjemite migmatite zone were primarily psammitic sediments whereas in the potassium migmatite zone pelites with aluminium excess for calcium and alkalies. The difference between the two belts has been interpreted to indicate two sedimentary basins connected with two different island arcs. The relatively young age of metamorphism in the potassium migmatite zone is due to a magmatic underplating after the collision between the Central and Southern Finland Island Arc Complexes later than 1.888 Ga ago. The volcanic rocks have same ages in both arcs indicating that the throndhjemite migmatite zone may represent an accretionary prism between the two arcs within which is situated also the suture zone of the two island arc complexes [20].

The deep conductor in southern Finland follows the throndhjemite zone and does not extend into the potassium migmatite zone. The conductor is exposed in the west but has no surface expression in the east, towards the Archaean-Proterozoic boundary zone. Near-surface conducting rocks, similar to those in the west but without any deep expression have been imaged north of the deep conductor in the east. The geometry of the conductors may indicate that deeper in the crust the collisional boundary between the two arc complex is E-W directed and follows the southern part of the throndhjemite migmatite belt to the east. Thus the southern part of the throndhjemite migmatite belt in the east is covered by rocks overthrust from south to north. The geometry and position of the deep conductor in southern Finland indicate that the conductor is situated in the collision zone of two tectono-metamorphic belts.

5. Conclusions

Based on geophysical and geological observations the following conclusions have been made:

I. The thickening of the crust and the high heat flow in the boundary zone between the Archaean continent and the Palaeoproterozoic Svecofennian Island Arc Complex was caused by magmatic underplating 1.885 Ga ago. II. The thick crust was stabilized isostatically soon after magmatic pulses dated at 1.885 Ga and has been preserved since then due to its increased density. The entire Svecofennian crust reached nearly the isostatic equilibrium between 1.885 and 1.810 Ma ago due to magmatic underplating. Strong thinning of the crust around the intrusions of the rapakivi granites is only local. III. Crustal thickness variations in the Archaean Domain, close to the Archaean-Proterozoic boundary zone, are also related to the magmatic underplating that caused the Svecofennian overprinting in the Archaean. IV. The geometry and position of the deep conductor in southern Finland indicate a collision zone between the two tectono-metamorphic belts. Geologically the collision zone has been interpreted as a suture zone between the Central and Southern Finland Island Arc Complexes.

Acknowledgements

The manuscript benefited greatly from the work of the GGT/SVEKA Working Group the results of which will be published elsewhere by the present authors and the GGT/SVEKA Working Group.

References

[1] Kontinen, A., 1987. An Early Proterozoic ophiolite - the Jormua mafic-ultramafic complex, northeastern Finland, *Precambrian Res.*, **35**, 313-341.

[2] Huhma, H., 1986. Sm-Nd, U-Pb and Pb-Pb isotopic evidence for the origin of the Early Proterozoic Svecokarelian crust in Finland. *Geol. Surv. Finland, Bull.*, **337**, 52 p.

[3] Korja, A. and Heikkinen, P., 1995. Proterozoic extensional tectonics of the central Fennoscandian Shield: Results from the Baltic and Bothnian Echoes from the Lithosphere experiment. *Tectonics*, **14**, 504-517.

[4] Luosto, U., 1991. Moho depth map of the Fennoscandian Shield based on seismic refraction data. In: H. Korhonen and A. Lipponen (eds.), Structure and Dynamics of the fennoscandian Lithosphere. Proceedings of the Second Workshop on Investigation of the Lithosphere in the Fennoscandian Shield by Seismological Methods, Helsinki, May 14-18, 1990. Institute of Seismology, University of Helsinki, Report S-25, 43-49.

[5] Korja, A., Korja, T., Luosto, U. and Heikkinen, P., 1993. Seismic and geoelectric evidence for collisional and extensional events in the Fennoscandian Shield - implications for Precambrian crustal evolution. *Tectonophysics*, **219**, 129-152.

[6] Heikkinen, P., Luosto, U., Malaska, J., Yliniemi, J. and Komminaho, K., 1996. Seismic structure along the GGT/SVEKA transect. (manuscript in preparation)

[7] Korja, T. and Hjelt, S.-E., 1993. Electromagnetic studies in the Fennoscandian Shield - electrical conductivity of the Precambrian crust. *Phys. Earth Planet. Int.*, **81**, 107-138.

[8] Rasmussen, T.M., Roberts, R.G. and Pedersen, L.B., 1987. Magnetotellurics along the Fennoscandian Long Range Profile, *Geophys. J. R. Astr. Soc.*, **89**, 799-820.

[9] FENNIA Working Group, 1996. P- and S-velocity structure of the Fennoscandian Shield beneath the FENNIA profile in southern Finland. *Annales Geophysicae, Supplement I to Volume 14*, C64.

[10] Ali Riahi, M. and Lund, C.-E., 1994. Two-dimensional modelling and interpretation of seismic wide-angle data from the western Gulf of Bothnia. *Tectonophysics*, **239**, 149-164.

[11] Ostrovsky, A.A., Flueh, E.R. and Luosto, U., 1994. Deep seismic structure of the Earth's crust along the Baltic Sea profile. *Tectonophysics*, **233**, 279-292.

[12] Chekunov, A.V., Starostenko, V.I., Krasovsky, S.S., Kutas, R.I., Orovetsky, Yu.P., Pashkevich, I.K., Tripolsky, A.A., Eliseev, S.V., Kuprienko, P. Ya., Mitrofanov, F.P., Sharov, N.V., Zagorodny, V.G., Glaznev, V.N., Garetsky, R.G., Karataev, G.I., Aksamentova, N.V., Guterkh, A., Grabovska, T., Koblanski, A., Ryka, V., Dadlez, R., Tsvoidzinsky, S., Korhonen, H., Luosto, U., Gaal, G., Zhuravlev, V.A., Sadov, A.S., 1993. Geotransect Euro-3 (EU-3). *Geophysical Journal, Ukrainian Academy of Sciences*, **15**, No. 2, 3-31, (in Russia).

[13] Sharov, N.V., 1993. The Structure of Lithosphere of the Baltic Shield. Russian Academy of Sciences, National Geophysical Committee, Moscow, 166 pp.

[14] Matthews, P.A. , Graham, D.P. and Long, R.E., 1992. Crustal structure beneath the BABEL Line 6 from wide-angle and normal-incidence reflection data. In: Meissner, R., Snyder, D., Balling, N. and Staroste, E. (eds.), The BABEL Project, First Status Report, Deep Reservoir Geology programme, Commission of the European Communities,Directorate-General Science, Research and Development, Brussels, Belgium, 101-104.

[15] Pajunpää, K., 1987. Conductivity anomalies in the Baltic Shield in Finland, *Geophys. J. R. Astr. Soc.,* **91**, 657-666.

[16] Koistinen, T.J., 1981. Structural evolution of an Early Proterozoic strata-bound Cu-Co-Zn deposit, Outokumpu, *Finland. Trans. Royal Soc. Edinburgh, Earth Sci.,* **72**, 115-158.

[17] Korsman, K., Hölttä, P., Hautala, T. and Wasenius, P., 1984. Metamorphism as an indicator of evolution and structure of the crust in eastern Finland. *Geol. Surv. Finland., Bull.,* **328**, 40 p.

[18] Hölttä, P., 1988. Metamorphic zones and the evolution of granulite grade metamorphism in the Early Proterozoic Pielavesi area, central Finland, *Geol. Surv. Finland, Bull.,* **344**, 50 p.

[19] Elo, S., 1994. Preliminary notes on gravity anomalies and mafic magmatism in Finland. In: Pajunen, M. (ed.), High temperature - low pressure metamorphism and deep crustal structures. *Geol. Surv. Finland, Guide 37,* 49-55.

[20] Lahtinen, R., 1994. Crustal evolution of the Svecofennian and Karelian domains during 2.1-1.79 Ga, with special emphasis on the geochemistry and origin of 1.93-1.92 Ga gneissic tonalites and associated supracrustal rocks in the Rautalampi area, central Finland. *Geol. Surv. Finland, Bull,* **378**, 128 p.

[21] Elo, S., 1984. On the gravity anomalies and the structure of the Earth's crust along the SVEKA-profiles. Report Q18/21/84/1, Department of Geophysics, Geological Survey of Finland, 5 + 17 p.

[22] Kontinen, A, Paavola, J. and Lukkarinen, H., 1992. K-Ar ages of hornblende and biotite from Late Archaean rocks of eastern Finland - interpretation and discussion of tectonic implications. *Geol. Surv. Finland, Bull.*, **365**, 31 p.

[23] Korja, T., Luosto, U., Korsman, K., and Pajunen, M., 1994. Geophysical and metamorphic features of the Palaeoproterozoic Svecofennian orogeny and Palaeoproterozoic overprinting on Archaean crust. In: Pajunen, M. (ed.), High temperature - low pressure metamorphism and deep crustal structures. *Geol. Surv. Finland, Guide 37*, 11-20.

[24] Kilpeläinen, T., Korikovsky, S., Korsman, K. and Nironen, M., 1994. Tectono-metamorphic evolution in the Tampere-Vammala area. In: Pajunen, M. (ed.), High temperature - low pressure metamorphism and deep crustal structures. *Geol. Surv. Finland, Guide 37*, 27-34.

[25] Väisänen, M., Hölttä, P., Rastas, J. , Korja, A. and Heikkinen, P., 1994. Deformation, metamorphism and the deep structure of the crust in the Turku area, southwestern Finland. In: Pajunen, M. (ed.), High temperature - low pressure metamorphism and deep crustal structures. *Geol. Surv. Finland, Guide 37*, 35-41.

Proc. 30th Int'l. Geol. Congr., Vol. 4, pp. 91-99
Hong Dawei (Ed)
© VSP 1997

GLOBAL GEOSCIENCE TRANSECT ACROSS EURASIA-RUSSIAN PART (from Murmansk to Altai)

Egorov A.S., Terentiev V.M., Bulin N.K., Kropachev A.P., Roose V.V., (VSEGEI, RUSSIA), Egorkin A.V., Kostychenko S.L., Solodilov L.N., (GEON, RUSSIA).

Deep modelling of lithosphere along global geotransects (GGT) is one of principal lines of the Russian Regional Investigational Programme.

All Russian Geological Research Institute (VSEGEI) in cooperation with GEON Centre are carrying out comprehensive interpretation of the geological-geophysical data on five geotransects and among them on Russian part of the International Murmansk-Taiwan GGT (8000 linear km in extent).

Cartographic design and digitization of the transect have been accomplished in terms of GGT Programme in cooperation with the Chinese team of scientists headed by prof. Yuan Xuecheng.

The Russian fragment of the transect crosses geostructures of different age of consolidation: Karelian-Svecofenian East-European Platform; Baikalian Barents Platform; Hercynian Uralian fold area; young West-Siberian Basin on the heterogeneous Caledonian-Hercynian basement; Caledonian Altai-Sayan fold area (Fig.1).

The foundation of modelling is informational data base involving:

- geological mapping data,
- multiwave deep seismic sounding data,
- gravimetrical data,
- magnetometrical data,
- magnetotelluric data,
- geothermal data,
- deep and superdeep borehole data,
- rock petrophysical data.
- remote surveys and lineament analysis of topographical data.

Vast informational data base and scientific and technical problems solving in the course of investigations required the development and application of an adequate methodology of volumetrical geological-geophysical modelling. It includes the set of interconnected procedures of processing of geological and geophysical data.

The key units of it while compiling geological-geophysical maps on different deep levels are the procedures of filtering, consecutive reduction of gravimetric and magnetic fields and commutation of their spectrum correlated components.

A wide set of the procedures of primal and inverse gravimetrical and magnetometrical problems is used for the vertical section modelling. DSS data are processed by means of statistical procedure which provides the seismic section presentation in the heterogeneity indicators and by these means reflects flat dipping boundaries. Due to the geothermal modelling the lithosphere thickness is estimated (Fig.2).

The choice of scientifically grounded theoretical model of the lithosphere deep structure is of great importance as it determines the contents of deep maps ,sections and their legends.In the course of our investigations we were guided by considerations of the Plate-tectonic theory taking into consideration resolution of the geophisical methods used.

92

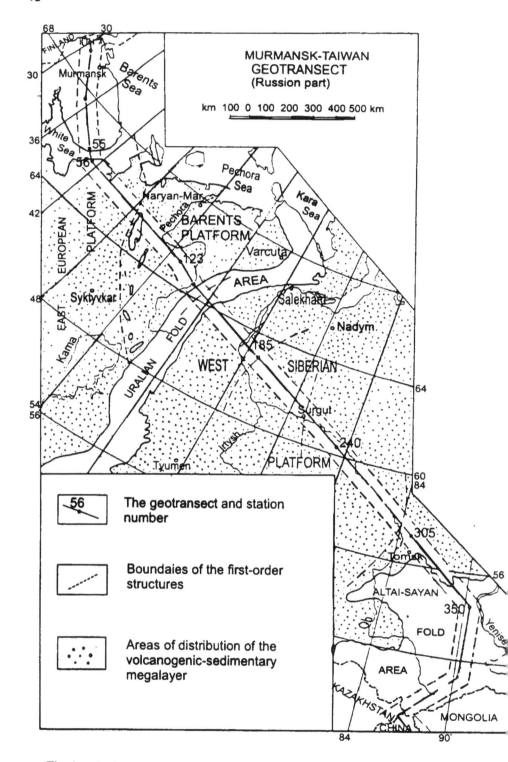

Fig.1.   Index map of the Murmansk-Taiwan transect (Russian part).

GLOBAL GEOSCIENCE TRANSECT (GGT)

(RUSSIAN PART OF THE RUSSIAN - CHINESE MURMANSK - TAIWAN GGT)

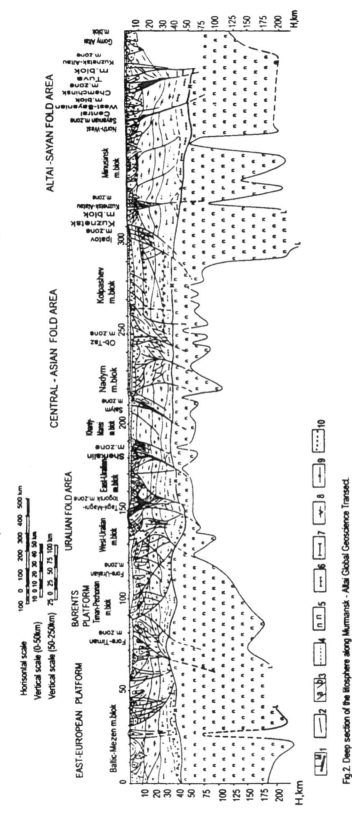

Fig.2. Deep section of the lithosphere along Murmansk - Altai Global Geoscience Transect.

1 - regionally traced geophysical bases of the Earth's crust (M-Moho discontinuti) and of the geotermal the lithosphere (L); 2 - boundaries of seismic and geological stratification of the Earth's crust; 3 - faults (a-major,b-others); 4 - uncertain traced geological - geophysical bundaries.
Structural petrological complexes of the litosphere : 5 - upper mantle ultrabasites; 6,7 - basites, basit-granulites of the mantle-crust mix (6) and of the lower crust (7); 8 - gneise-basit-granulites of the middle crust; 9 - gneises of the upper crust; 10 - subhorisontal low velocity zones. (Symbols of structural - petrological complexe of the volcaonogenic - sedimentary folded layer and slightly dislocated megalayer are as Fig.3).

The principal subjects of deep geological-geophysical modelling are massifs (megablocks) of ancient continental crust and interblock megazones corresponding to variety of extensional, compressional and shear environments.

Deep section of the ancient continental "normally stratified" lithosphere includes (downwards) volcanogenic-sedimentary megalayer, three megalayers of the consolidated Earth's crust (upper crust Vp=6.0-6.4km/s ,middle crust Vp=6.5-6.8km/s,lower crust Vp=6.9-7.2km/s), upper mantle megalayer Vp>7.8km/s.

Interblock megazones are characterized by seismically unstratified crust. Their internal structure and petrological composition are estimated from results of gravimetrical and magnetometrical modelling.

We place of major emphasision on such anomalous geophysical objects as: lower crust areas with ivcreased over 7.2 km/s velocity ;upper mantle areas with anomally low velocity (<7.8km/s); crust and upper mantle subhorizontal low velocity zones.

Rock associanions are classified according to age and geodynamic environments of their formation.

The major features of the lithosphere dssp section along the geotransect are as follows (Fig 2).

The basement of East - European Platform is tectonically the geostructure of the Early-Proterozoic collisional accretion of ancient Archean lithosphere segments of different composition and denudation level.

It is characterized by increased lithosphere thickness (200-250 km), crust thickness varies from 30 to 43 km. Ancient crust segments are characterized by thin stratification. Sharp change of the deep section structure observed in Fore-Timan megazone which dips south-west and extends up to the lithosphere base. In the crust it is modelled as the set of tectonic wedges consisting of the paleoocean crust complexes. To the east of the zone an orogenic belt with intense granitization is modelled.

The geostructure is disintegrated by a set of Late Archean-Early Proterozoic interblock riftogenic structures. The deep structure of Imandra-Varzuga riftogenic zone is formed by the set of normal faults which are confined to middle-low crust destruction zone. Low-crust area with increased Vp is adjacent to the thrown side of the zone. It is considered to be the newly generated transitional mantle-crust mix. It seems to reflect the results of the process of the crust building up.

The geostructure is characterized by an extensive development of thrusts and overthrust nappes. The majority of them bottom on low velocity zones in upper and partially in low crust.

Inside the Baltic Shield the crystallin basement comes out to the day surface, in Mezen syneclise it is buried under thick (3-4 km) Riphean-Phanerozoic sedimentary cover.

To the east the geotransect crosses the Barents Platform, which is considered to have been formed on the Riphean fold basement involving massifs of ancient continental crust.

Intermediate lithosphere thickness of the geostructure averages between 90 and 170 km.

There are three megastructures of different type in its basement.

Judging from geological-geophysical data the Timan megazone is defined as Riphean fold structure overthrusted on the East-European Platform edge. On evidence derived from gravimetrical data the area with increased density is modelled at 5-10 km depth beneath Riphean metaterrigenous complex. This anomaly might be caused by the complexes of the Early Riphean rift which might have developed up to intercontinental stage. Low crust rift pillow is modelled in thrown side of this zone.

Timan-Pechorian megablock is the ancient Precambrian continental massif characterised by normally stratified crust covered by 5-8 km Riphean sedimentary-metamorphic complex and 3-4 km Phanerozoic sediments.

Fore-Uralian megazone, partially covered by Uralian tectonic nappe, is characterized by anomalous geophysical manifestation: disturbed seismic stratification ;with no granitic layer crust with high velocity (6.6-6.65 km/s) complexes on the upper crust level. We speculate island arc origine of this megastructure.

Extensivelly developed thrust dislocations bottom on subhorizontal upper and low crust low velocity zones and form imbricated-thrust structures.

Uralian overthrust-fold assemblage is interpreted to be a bivergent compressional orogen. In accordance with results of geophysical modelling the Uralian fold area is much wider than geologically mapped Hercynian structures and consists of megablocks with continental crust and suture megazones separating them. It is characterized by thick crust (to 56 km) and thin thermal lithosphere (60-100 km).

West-Uralian tectonic nappe is built up on the Barents Platform edge. Tectonic wedges containe complexes of different composition and age from Precambrian to Ordovician-Permian Hercynides. Preduralian foredeep is located on the west edge of Uralian fold area and is formed from Permian-Triassic molasse.

Tagil-Magnitogorsk suture megazone features bivergent structure of fault deformations with ophyolitic and island arc allochtons overthrusted the neighbouring continental megablocks. The bowl-shaped ophyolitic body is modelled at the depth of 5-12 km. The set of the fold-thrust deformations bottoms on subhorizontal zones traced at 10 km. The problem of tracing of the suture megazone at the lower crust levels is under discussion. We assumed its west dipping having regard to the geophysical data interpretation.

West part of East-Uralian continental megablock contains the collisional belt with a set of deep through-crust deformations and intense granite generation. The Hercynian basement is covered by 2-3km thick sediments.

Sherkaline megazone is similar in deep structure and composition to Tagil-Magnitogorsk megazone.

The basement of young West-Siberian Platform is defined as composite tectonic collage formed in the course of Paleozoic continental accretion during the closure of Uralian and Asian paleooceans [4]. Linear structures of the Uralian fold area are separated from the mozaic one of the Central Asian fold belt by Kazakhstan composite continent. The lithosphere thickness varies from 50 to 100 km inside the West Siberian Platform.

Khanty-Mansi and Nadym continental megablocks were accreted and incorporated into composite Kazakhstan continent in Early Paleozoic time. They are separated by the Salym megazone flatly dipping westwards. Within the later a set of tectonic wedges of ophyolitic and island arcs complexes is modelled.

In the basement of Central part of West-Siberian Platform the relict depression formed on the ocean crust is located [1]. Judging by geophysical data it is composed of thick Paleozoic sedimentary and volcanogenic complexes.

Kolposhev continental megablock is in the Eastern partof West-Siberian Platform and is bounded on the west by Ipatov megazone flatly dipping westwards and marking the deep boundary of the West Siberian and Altai-Sayan geostructures by sharp increase of lithosphere thickness from 75-100 km to 150-200 km.

Intracontinental rifts composed of Triassic volcanogenic-terrigenous complex entrenched into the Paleozoic basement of the West-Siberian Platform. There are characterized by trough-like structures with relatively narrow root zones which manifest themselves by desintegration of the crust seismic statification . Along the transect the overage thickness of the Mezozoic terrigenouscarbonate cover is 3-4 km.

Altai-Sayan fold area is characterized by an increased lithosphere thickness (to 150-200 km) and crust (to 52-53 km).The characteristic feature of it is an extensive spatial development of the crust-mantle mix.

On the west the geotransect crosses Kuznetsk and Minusinsk continental megablocks separated by Kuznetsk-Alatau megazone. This extensive tectonic structure (several thousands kilometres in extend) represents deep through-lithosphere destruction zone. At the subsurface depth it manifests itself as the assemblage of major strice-slip faults surraunded by normal and thrust faults. In vertical section the assemblage is configured as "flower structure".

The characteristic features of Minusinsk megablock deep structure along the transect are active Paleozoic granitization and wide development of overthrust and thrust dislocations .The later botom on the subhorizontal zone of desintergation located along the root of the upper-crust "granitic-gneiss" megalayer. Judging from geophysical data a set of deep flat foults penetrates through middle and low crust.

Central West-Sayanian megablock having anomalous continental crust is bounded on the north and south sides by north vergent suture zones. The later contain tectonic wedges of ophyolitic and island arc complexes. Upper structural stratas of the geoblock are composed of Riphean-Middle-Paleozoic sedimentary complexes. They are intensely intruded by collisional granitoids.

Then the geotransect crosses Tuva megablock having very intensely granitized Archean-Proterozoic basement, futher again crosses Kuznetsk-Alatau strike-slip megazone and passes into Gorny Altai megablock. The characteristic feature of the latter in parallel with intense granitization is a wide development of volcano-magmatic arc complexes.

The geological-geophysical modellings fulfilled along the geotransect allowed to draw some regular trends in deep structure of the crust adding the starting scientific-theoretical model:

- up to 80% of the Earth's crust is composed of massifs (plates, microplates, segments) with ancient continental crust;
- sets of overthrusts, strike-slip and normal faults are genetically and spatially correlated with interblock megazones;
- sharp structural disharmony among Earth's crust strata separated by subhorizontal destruction zones (low velocity zones) is noted.

Geological-geophysical modellings along geotransect are supplemented by plate-tectonic reconstructions of the deep crust structure to the major stages of their geological history fulfilled with the use of the global paleoreconstructions [2,3,4]. Sketch paleo-sections in combination with space-time diagram (Fig 3) show the set of ancient crust massifs (paleo-plates, microplates), sequence, age interval, geodynamic environments of the formation of the principal Proterozoic and Paleozoic geological complexes and structures and the appropriate complication of the deep section

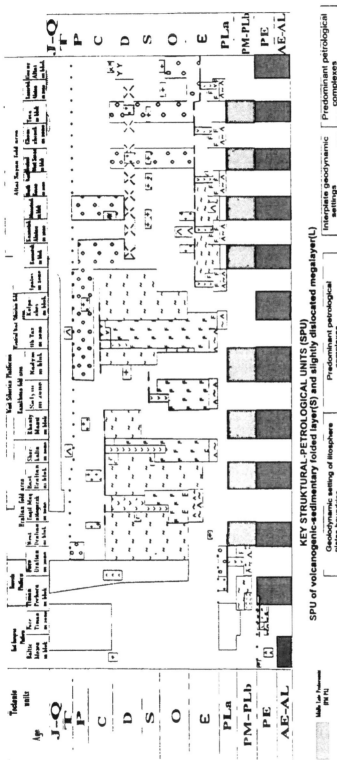

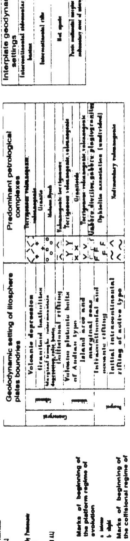

KEY STRUKTURAL-PETROLOGICAL UNITS (SPU)

SPU of volcanogenic-sedimentary folded layer(S) and slightly dislocated megalayer(L)

Fig.3. Space - time diagrame

of the Eurasia in the course of the Paleozoic tectonic-magmatic cycle. Besides, some ideas concerning characteristic features of the Proterozoic evolution of the continent are summed up in schematic cartographical form too. For example, set of paleo-sections shows the effect of the East-European continental crust destruction and thinning and trough zones emplacement in the course of Karelian riftogenic stage. In the closing -collisional stage orogenic belt of the Fore-Timan megazone was formed, thickness of the East-European Platform was increased due to mantle-crust mix and thrust-nappe deformations on different deeps of the crust were formed.

The Timan rift was emplaced during Early Riphean. This structure might have developed up to intercontinental stage. At the convergence stage the thick island arc was emplaced along the Barents plate boundary. Later the arc was accreted to its east edge.

Formation of the Uralian branch of the Paleozoic Uralian-Mongolian fold area was being continued from Early Cambrian to Permian.

Opening of the Uralian paleoocean (sea-floor spreading) began in Early Cambrian and continued up to Ordovician when its width ranged up to two thousands kilometers [4].

In Silurian Khanty-Mansi and Nadym microplates collided with Kazakhstan continent and with one another having formed Salym and other fold structures. In this interval along the eastern boundaries of the continental plates and microplates island arcs were emplaced.

By the Middle Carbonianiferous the whole of the Uralian paleoocean crust had been fed into trenches and collisional processes started producing thrust-nappe structures and intense granitization. We interpret the intense granitization of the west edges of the East Uralian and Khanty-Mansi megablocks as additional evidence of westward dipping of the Tagil-Magnitogorsk and Sherkalin sutures. Closing of the Sherkalin branch of the Uralian paleoocean took place later than closing of the Tagil-Magnitogorsk branch and that feature determined complication of the Tagil-Magnitogorsk megazone by west vergent thrusts. Uralian tectonic deformations had been active for a long time and, in particularly, deformed Permian molasse deposits of the Fore-Uralian foredeep.

The evolution of the Central-Asian branch of the Paleozoic Uralian-Mongolian Fold Belt was proceeding in a different way.

Opening of the Asian paleoocean took place in Vendian and by Early Cambrian it reached maximal width (3000 km). In Early Cambrian along margins of numerous continental microplates the set of island arcs was formed. The collision of the Siberian continent with numerous microplates and island arcs of the Altai-Sayan region took place in Late Cambrian. Along and across suture zones thrust-nappe and strike -slip dislocations were emplaced following the collision. The latter were predominant in this region and active during the entire Paleozoic.

During the Late Cambrian-Ordovician a great number of granitoid plutons was intruded into the colliding blocks. In some structural environments of the Central West-Sayanian and Tuva and some others megablocks flysch and molasse complexes were developed. At this time the vast Asian ocean basin and some relict interior basins with oceanic crust in the Altai Sayan region remained.

During the Silurian an additional phase of granitoid magmatism took place caused by closure of the relict interior ocean basins.

During the Devonian the Asian paleoocean width greatly decreased. Inside the Altai-Sayan fold area in parallel with sedimentation formation of volcanic depressions of discussed origine took place. Within Gorny Altai the calc-alkaline volcanism is associated with Hercynian subduction processes at the region boundaries.

The Carboniferous is the time when the Asian paleoocean was closed, although separate relict basins with ocean crust remained. In depressions in outlying districts of the Altai-Sayan fold area molasse complexes were deposited.

By Permian the island arc activity had been completed and since the Late Carbonian in relict basins thick molasse complexes had been depositing.

During Late Permian-Early Triassic the complex deep structure of the north area of the Uralian-Mongolian Fold Belt emerging from the Paleozoic evolution was disrupted by intracontinental rifting along zones of weakness. As a consequence, West-Siberian area suffered dipping which led to formation of the Mezozoic sedimentary basin.

CONCLUSIONS:

1. The set of maps and sections along the Russian fragment of the International Murmansk-Taiwan geotransect has been accomplished in terms of GGT Program with the use of the original methodological scheme of interpretation of the vast geophysical data base and in the framework of the scientific theoretical model hose parameters have been developed on the basis of the Plate Tectonic theory taking into consideration the resolution of the geophysical methods used.

2. The procedures of paleoreconstructions of the deep section permitted to visualize stages of plate-tectonic evolution of the Eurasian continent along the transect , to make corrections in the initial version of the deep section and thus to obtain the final version which did not contradict the data base used and the authors' ideas of the history of the Eurasian continent evolution.

3. The results obtained can be used as a serious geological basis for solving of a wide specrum of scientific and applied problems (deep geological-geophysical mapping under thick sedimentary cover, predictive metallogenic, geoecological and other investigations).

REFERENCES

1. Aplonov S.V. The tectonic evolution of the West Siberia: an attempt at geophysical analysis. Tectonophisics, 245 (1995) p.p. 61-84 (In English).
2. Geodinamical map of the USSR and adjusted acvatories (on 1:2500000 scale) Editors: Zonenshine L.P., Meselovsky N.V., Natapov L.M. Moscow, 164 p. (in Russian).
3. A. Ronov, V. Khain, K. Seslavinsky. Atlas of lithological-paleogeographical maps of the Word. Late Precembrian and Paleozoic of continents. Leningrad. 1984 (in Russian).
4. Zonenshine L.P.A History of the Uralian Ocean. Nauka, Moscow, 164 p. (in Russian).

Proc. 30* Int'l. Geol. Congr., Vol. 4, pp. 101-117
Hong Dawei (Ed)
© VSP 1997

The Lithospheric Textural and Structural Features as well as Oil and Gas Evaluation in Lower Yangtze Area and Its Adjacent Region, China.

Chen Husheng, Zhang Yonghong. Xu Shiwen. Yan Jizhu. Guo Nianfa. Zhu Xuan
(East China Bureau of Petroleum Geology, MGMR, China)

Abstract

The work presented in this paper is based on the results from deep geology investigation project(85-06-203,1991-1995) sponsored by Chinese ministry of Geology and Mineral Resources (MGMR).From the research and investigation, five significant discoveries were made. They are 1.the discovery of Dabie-Zhoushan lithospheric fault and Binhai-Tonglu lithospheric cryptofault.2.the discovery of the effection of expanding diapirism in deep geology.3 the discovery of existence of at least three horizontal fracture zones in lithosphere.4 The discovery of low resistivity layers under metamorphic rocks (nappe), which might be some unmetamorphic sediments or relics of horizontal fracture zones after uplifting, they both indicated new oil and gas exploration domains and 5.The finding of marine strata or unmetamorphic deposit covered by volcanic rocks in eastern Zhijiang Province, it is also a new domain for hydrocarbon survey. In working area, the structures of lithosphere were studied and they could be classified into seven layers and eight blocks. Three rules which controlled uplift and subsidence blocks movement were suggested, and to the movement mechanism, a continent dynamic pattern with four comprehensive factors was established. The interpretation on three sets of fault systems for the first time make contribution to understand structures in the area. In this paper, the evolution of lithosphere was divided into five phases. The hydrocarbon resources were evaluated, and the next steps in oil and gas exploration are proposed in eight aspects.

Key Words: Significant discoveries, structure in layers and blocks, movement rules of uplift and subsidence blocks, three fault systems. The evolution of lithosphere, hydrocarbon evaluation

The work presented in this paper is based on the results from deep geology investigation project (85-06-203,1991-1995) supported by MGMR. The project includes three subjects and thirteen monographs. Under the guideness of modern geology theory. particularly in the advance of plate tectonics. the research was performed systematically. it involved in a great deal of geophysical and geologic data measured in this area. they include 819 magnetotelluric sounding points, 819.16KM crust depth reflection profiles(record length:12S), $5.5*10^4$ Km seismic data (record length:6 S), 75 samples for high temperature and high pressure experiment, 487 specimen from 14 layers for paleomagnetism,$47.5*10^4$ Km^2 gravity data. $35*10^4$ Km^2 airborne magnetic data. and the data from more than 2000 wells in various depth as well as other new data in this area (Fig.1). Consequentially the main research results arc as follows.

1. Significant discoveries

(1) Discovery of Dabie-Zhoushan lithospheric fault and Binhai-Tonglu cryptofault in lithospheric level for the first time (Fig.2).

(2) Discovery of expanding diapirism of basement. Because of impulsive expansion of earth and regional expansion. lithosphere was split at bottom first. then the fracture gradually extended to the surface. to form a vertical fractural zones originated from lithospheric. It is caused to compose a conjugate fracture network in the region. The knowledge of this phenomena would be very important in geology (Fig.3).

(3) At least three sets of horizontal fracture zones were found in lithosphere. In geoscience profiling it indicated that some layers with high conductivity and low velocity always existed in this region. To the high conductivity layer only. seven to eight interpretation could be available. but to the layer with low velocity and high conductivity simultaneously, just few cause such as rupture or liquidation was suitable. According to the result of the experiments in high temperature and high pressure. rupture of various rocks was occurred only within three different depth limit, and at the same time these layers in low velocities. So the interpretation was that the layers either within these three depth limit. or with high conductivity and low velocity were three fracture zones filled by liquid respectively, more over. the low velocity layer within the depth from 15-22 Km. was defined as first horizontal fracture zone. it divided the crust into upper and lower parts. the second zone was located within 33-36 Km in depth. which was correspond to Moho layer in this region. the third horizontal fracture zone was the low velocity layer with 50-60 Km in depth and the upper and lower mantle was separated by this zone (Fig. 4,5).

(4) Discovery of low resistivity layer under metamorphic rocks(nappe).which might be some Meso-Paleozoic marine sediments that is an important domain for hydrocarbon exploration, or the relics of uplifting horizontal fracture zones which would be good reservoir for inorganic oil and gas pool. This potential area for inorganic oil and gas survey meits our attention.(Fig.6).

(5) Discovery of marine strata or unmetamorphic rocks under volcanic rocks in east Zhijiang province is another important achievement. it also extends the domain for hydrocarbon exploration (Fig.6).

2. The Lithospheric texture in layers and block

Based on the HQ-13 line data and the lithospheric investigation in this region. it is demonstrated that the lithosphere was constructed by texture in layers and blocks. i.e. separated by layers in vertical and divided by blocks in horizontal [1]. According to present techniques and existing data. the lithosphere in this region can be divided into 7 planes and 7 layers in vertical. and isopach map and bathymetric map of all these planes and layers listed in the following table have been completed. The lithosphere can be divided into many blocks laterally. According to the concepts of modern structure unit division which can reflect the lithosphere textural feature, the blocks were divided into as follows(Fig.2)

(1)East China continental margin plate which consists of:

a: Lower Yangtze-basin region, comprising North Jiangsu basin and South Jiangsu uplift.

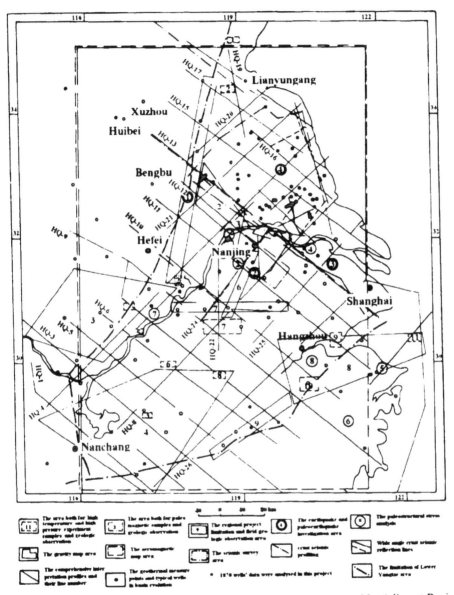

Figure 1. The Research Data Distribution Map in Lower Yangtze Area and Its Adjacent Region Lithospheric Project.

b: Shangdong-Jiangsu uplift region
(2) South China plate, which consists of:
a: Dabie uplift region
b: Wangjiang-Poyang rift system
c: Jiangnan-Qiantang uplift region
d: Jiangshan-Shaoxin rift system

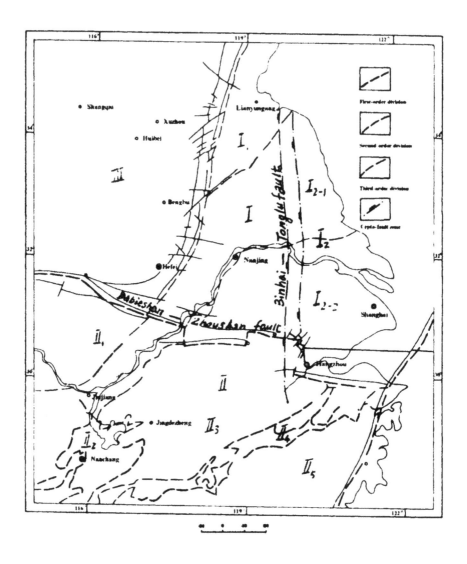

Figure 2. The structural division in Lower Yangtze area.

I_1 : Lusu Uplift. I_2 : Lower Yangtze rift basin. I_3 · North Jiangsu basin. I_4 : South Jiangsu Uplift. II_1 · Dabie Uplift. II_2 : Wangjiang-Poyang rift. II_3 : Jiangnan-Qiantang Uplift. II_4 : Jiangshao rift. II_5 : Cathaysian Uplift.

e: Cathaysian uplift region

(3) North China plate (only southeast part within this region, as a sub-unit)

This structure division principle is different from the traditional one. It is not dominated by the principle based on geologic historic development, but emphasizes a comprehensive result of lithospheric evolution from past to present day, stressing modern tectonic circumstances,(that is, middle-late Yangshan-Himalayan orogeny movement) . Establishing structural division map for Yangtze area is mainly based on modern lithospheric texture, gravitational field, magnetic field, geothermal field and

surface extension rift

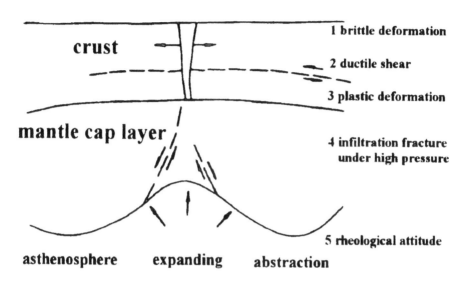

crust

mantle cap layer

1 brittle deformation

2 ductile shear

3 plastic deformation

4 infiltration fracture
under high pressure

5 rheological attitude

asthenosphere   expanding   abstraction

Figure 3. The sketch of base expansion rift.

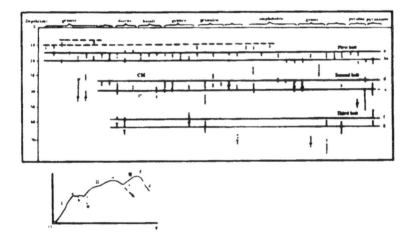

Figure 4. Rock testing under the condition of high T-P.

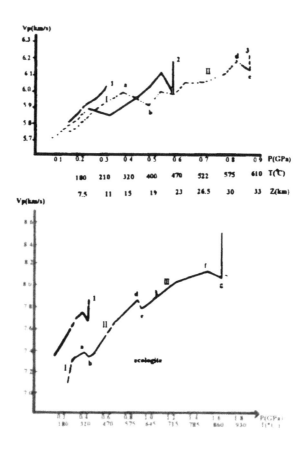

Figure 5. Part of the original recording of granite and ecologite.

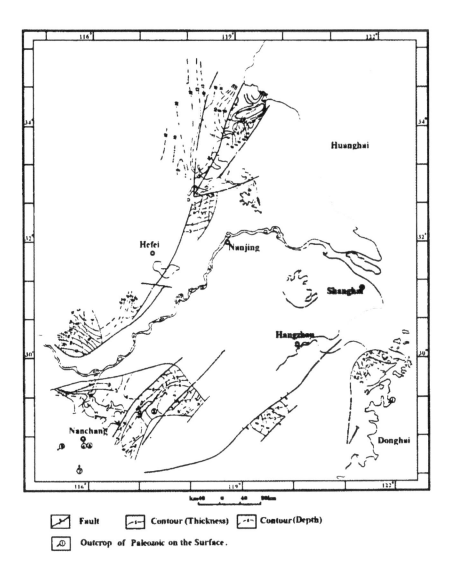

Figure 6. Nappe in metamorphic rocks and frontier distribution of marine Mesozoic and Paleozoic Group in Lower Yangtze and Its Adjacent Area.

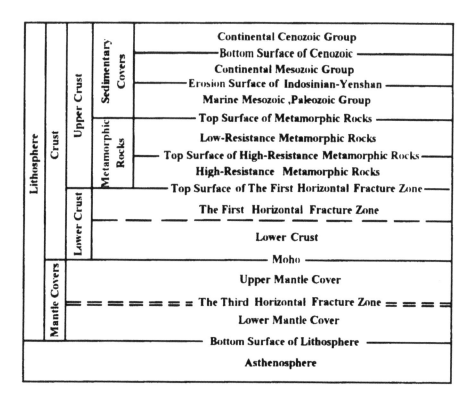

Table 1. Division Table of Lithosphere Layers in the Lower Yangtze and Its Adjacent Area.

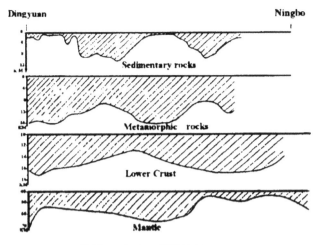

Thickness Section of HQ-11 Line

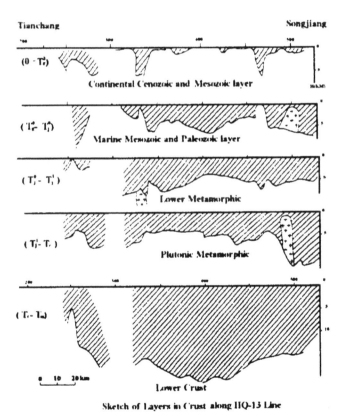

Sketch of Layers in Crust along HQ-13 Line

Figure 7. HQ-13 Seismic Sounding Crust Thickness

divergent plate-marginal fault. The uplift region belongs to mountain-thick crust-thick mantle type, but the subsidence region is characterized as plain-thin crust-thin mantle type and hot crust, hot mantle type, it can be used for petroleum and mineral resource exploration, it also can forecast earthquake and reduce geological diasters, as well as provide useful data about the global changes in environment and climate. because all these fields are controlled by current structures. Now it is summarized the lithosphere texture in lower Yangtze area and its adjacent region as "seven layers and eight blocks". This project which covers so wide area in lithosphere structures and division research really is an important event in deep geology investigation in China.

3. The movement laws of uplift and subsidence blocks structures (or basin and ridge structure)

The uplift and subsidence blocks are the basic structure unit in continental lithosphere. the subsidence block refers the negative structure in general, and uplift blocks is positive one. The movement laws of uplift and subsidence blocks appear wave motion phenomena.[2,3] These laws describe as follows:

a. Uplift is interlaced by subsidence. They are in associated state. This is the law about their space distribution. This features is reflected by various structure maps. it also can be seen in Fig.2

b. Uplift can be substitute with subsidence and transformed into subsidence. and vice versa. This is the law about their time evolution.

For example, the transform from negative area to orogenic zone would be the most active change for uplift and subsidence blocks.(the same as "return effection" in geosyncline theory). This kind of transform is not only happened in active area. but also in relative stable zones, some changes, such as oscillation. seesaw-like movement. smaller underwater transform and so on. are all obeyed this law, only smaller both in scale and level. This law also can be found in Fig.7.

c. The uplift and subsidence were migrated in particular time and direction. This is the law about their time-space association. There are at least three ways in the directional migration for uplift and subsidence blocks.

(1) Migration towards ocean. when the continental lithosphere was moved in stretch movement. the migration towards ocean occurred. Several large uplift and subsidence structural zone in middle and east China(i.e. Neocathaysian tectonic system named by Professor Li Siguang) from west to east were formed and developed in the order of time. from old to young. This is a typical example in migration towards ocean.

(2) Migration towards continent. When the amalgamation collisions among plates (terrain) were happened. the migration towards continent would be occurred. A typical example is, when the Indo-Chinese Plate was meet and prograded towards Qinhai-Tibet Plate. the uplift and subsidence structural zones were generated. from south to north, they gradually became younger.

(3) Migration with rotation. The plate block was shifting accompanied with rotation at the same time. For example, when the lower Yangtze terrain was drifted from south hemisphere to north hemisphere, the terrain itself was rotated in counter clockwise. and the direction of associated uplift and subsidence structure zones were shifted in counter clockwise too. The age from old to young. the directions were moved along WE-NE-NNE-NNW in order.

From the analysis about these three laws, it can be seen that the movement of uplift and subsidence blocks is reflected as a kinds of wave motion, with the feature of elastic wave, including superimposition principle, interfere phenomena and directional propagation.

4. Discussion on the dynamic mechanism for movement of uplift and subsidence structure.

Based on our research work, there are mainly four factors which controlled the dynamic action for the movement. In summaries, it was guided by horizontal compressing force or Earth expanding forces, limited by the fracture zone of continental lithosphere both in horizontal and vertical, dominated by lava activity in mantle (part of molten material in asthenosphere) and controlled by gravity isostatic adjustment. Of cause, it is a comprehensive process and assumed as follows .

It is supposed that a big horizontal compress or earth expanding force acted on a stable block, to make the lithosphere broken, formed a net-like vertical fracture zone. After this action, the stable region became active. It is originated to generate uplift and subsidence structure. This is the guideness of dynamic action.

When the lithospheric vertical fracture zone was formed in the region, the accompanied horizontal fracture zone have existed before. They were intruded by mantle lava. The lava came into vertical fracture zone first, then expanded them rapidly and prograded to other cracks, gradually melted or partly melted the lower crust of this area, the melted lower crust was less density, it made the upper crust sunk to form a rift basin, moreover, the melted lower crust occupied more volume and generated a side pressure to the adjacent area, under the action of this side pressure, the adjacent area was raised to become a uplift.

At the same time, the light part of lava from lower crust and mantle transformed the multiple phase eruptive rock in order, it was associated with rift basin. The eruptive affection with multiple phase caused subrift-settle of the area on many occasions, a large rift basin was composed by the superimposition of chain of rift and depression basins.

The heavier part of lava in lower crust could not introduce into the upper crust because its high density, but it was squeezed in the adjacent horizontal fracture zone and invaded into the other fracture one by both the pressures from crust sink and mantle lava hamp. Because of this activity, the lower crust in this area became thinner, and began a active period for a large scale invasion of medium-acidic igneous rock in the adjacent area. At that time a large amount of additional squeezed igneous rock made the adjacent area uplift in large extent. During this uplift period, a thrust nappe process is occurred in the boundary between uplift area and basin, a variety of nappe structures were constructed.

The raised area was consisted of ridge zones and erosion area. The ridge became thinner and thinner, at that time the gravity isostatic adjustment made it raise again, the metamorphic complex in deep would be brought out to surface by uplift or thrust nappe process, the eroded material came into basin to add the load to basin, let it settle continuously

The process described above reflected a period in which uplift and subsidence structures were developed in associate state, and their activities were very strong.

If there was no interfere from another big horizontal compression, the continues settle of basin either sequeezed the heavier part of melted lava in lower crust into uplift, or made heavier part of the mantle lava into the adjacent horizontal fracture zone along the top of mantle in uplift area, then a fusion zone which separated the crust in uplift area and mantle in lithosphere was formed, it would be let the unmolten mantle in lithosphere sunk into the bottom of asthenosphere, because the mantle in lithosphere posses higher density, as soon as it separated with crust, it might be sunk, i.e. the deiamination. At that time the "mountain root" in the raised area disappeared, rapid uplift began, and in a short time, the uplift would be transformed into subsidence.

As mantle lava was closed to crust in this area, the temperature in lower crust became higher, the result was the same as lava invaded into lithospheric vertical fracture zone, led on a series of subsidences and formed new rift basins. The area no matter whether or no the vertical fractural zone did exist, the basin always existed. This is the extention of stretch effection, let the continent rift basin extented to sea basin and oceanic basin, the uplift and subsidence in continent might be enlarged to global continent ocean structural system.

In the process mentioned above, if a strong horizontal compression was acted, in basin the rock with lower density and plastic deformation would be compressed and folded into a new orogenic belt (i.e. 'return effection'), but in uplift area, the rock with higher density and brittle deformation, might be pressed to fault zone. This fault zone would develop to become a new basin area, and the new orogenic belt of course was a new uplift area, thus a large scale movement in which the uplift and subsidence were replaced and transformed by the other side was completed. This process appears frequently in geologic history, as for the process in smaller scale should be much more. In this dynamic process described above, control by gravity isostatic adjustment always worked. The depth, limit and output state of lava activity with various densities were controlled. The relative thickness of different layers were adjusted equally, the generation and scale of uplift and subsidence were handled, all by the gravity isostatic adjustment. It combined with geothermal (lava flow) to play the key role in comprehensive dynamic process, to make the movement advanced in described law.

After such comprehensive dynamic process in several times, multiple transformation between uplift and subsidence were fulfilled. Mantle lava lost a great deal of thermal energy and reflected less activity. The lava in vertical fracture zone and developed fault area was gotten cold and solidified, to block up the channels for their flow, then to make the formed structure of uplift and subsidence dull. Finally the structure was trended to stable. This adjustment flatted the surface to quasi-plain, as a new craton.

This comprehensive dynamic process is called 'continent dynamic pattern with comprehensive four factors

The features of this pattern are as follows:

(1) The four factors in the pattern all are founded on facts. They are demonstrated on a number of data. The basis of theory is reliable.

(2) It is reasonable to consider subsidence (basin made event) and uplift (orogenic event) simultaneously, because the structures of uplift and subsidence is accosslated in nature, so did the dynamic process, the research only for uplift or subsidence should be avoided.

(3) At least four subsidences and three uplift effections were interpreted comprehensively in the pattern. It was clearly described that the generation and evolution of structure in surface were controlled and effected by the geologic events in deep.

(4) Some interesting geologic problems, such as igneous rock ivansion nappe structure, deiamination, return effection and global continent-ocean structure system etc. were interpreted and discussed.

5. Fault system

The fault system is one of the most important structure in lithosphere. It is a basic factor for dividing the lithosphere into different blocks. In research area, the fault system was very complex. After research, three fault systems were suggested at the first time.

(1) When the lower Yangtze Plate originated at early Indo-Sinian and Yenshan stage collided with North China Plate, a thrust nappe fault system was constructed.

(2) At the middle and late stage of Yenshan in East China large area, the expanding diapirism of basement was caused by plate split movement, thus, a net-like vertical fracture zone in lithospheric level represented by Tan-Lu fault and Dabie-Zhoushan fault system were generated. These huge framework fault became the boundary of new split plate, and composed the second fault system in this area, called plate split fault system (Fig.2).

(3) Within the new split plate, in the north of Dabie-Zhoushan fault, large rift basin was found, in the south of it, rift belt with narrow band in shape was superimposed on the thrust nappe system. The third fault system-extension rift fault system was based on such structural movements (Fig.9). General speaking, the interpretation about three sets of fault systems made much more progress in understanding the structures in this area.

6. The Evolution of Lithosphere

The lithospheric evolution includes the convergence and divergence of basic blocks, the development and change of sediment covers as well as the eruption and invasion of igneous lava. The evolution history is trinity. The evolution procedure may be summarized as five periods:

(1) The periods in Archaeozoic Era at that time the craton was formed

At that time, a lot of silico-aluminate(or granitic rock) blocks scattered at the region near equator were gradually converged to form cratons in various size. In research area, the craton groups which were centered about the carton of South Yellow Sea were distributed, and formed the similar metamorphic rock systems to that in the bottom of upper crust. This period was the primary phase of lithosphere construction, called carton generation phase.

(2) The period in Proterozoic Era at that time Lower Yangtze Plate was formed

The continent blocks (or terrain) centered around carton would be convergened further. In this area, after Jiuling and other blocks amalgamated with the center, i.e. South Yellow Sea carton in Protorozoic Era, the amalgamated block was continued to merge Huaiyu and other terranes, the whole area got into continental accretion period. The independent Lower Yangtze Plate and Proterozoic Era mesocratic and leucocratic

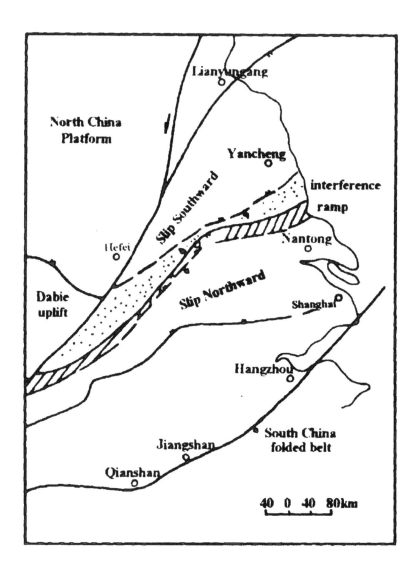

Figure 8. Ramp interference in south and north direction skeptically

metamorphic rock system were generated. This period is called Low Yangtze Plate generation phase.

(3) The period in Paleozoic Era at that time south old land was formed

At that time both Lower Yangtze Plate and Cathaysia blocks (or terranes) were drifted from south hemisphere to northeast. In Caledonian. Cathaysia continental block met

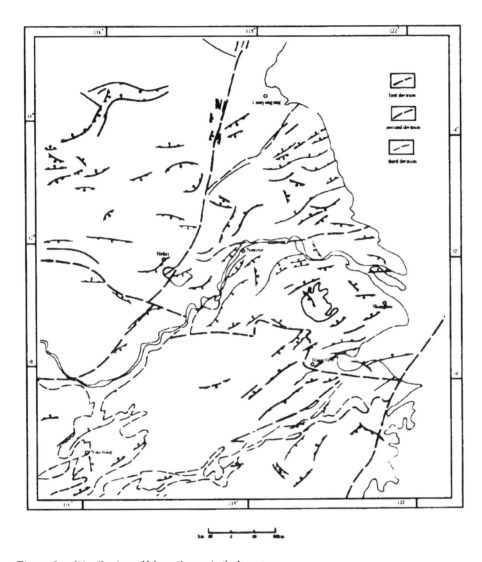

Figure 9. Distribution of Meso-Cenozoic fault system.

Yangtze Plate at first, then was amalgamated with them and the south old-land was composed. In uplift area, the raised part was folded and deformed by the amalgamation. In the north part of Upper Yangtze Plate, the amalgamation effection was not so strong that it still kept in marine deposit. This period was called as south old-land generation phase.

(4) The period at Mesozoic Era at that time the south continent was amalgamation with north one.

The south continent which included Upper, Middle and Lower Yangtze and Cathaysia subducted towards to North China Plate and then amalgamated with it.

The south and north continent were welded as a unique China continent by the Lian-

Huang subduction zone and Dabie-East Sandong collision mountain belt. The sediments from Later Paleozoic to Early and Middle Triassic Period were deposited in the meantime. After collision, the brine secede overall. From then on, Lower Yangtze area was entered in the non-marine basin development stage. This period was called south and north continent amalgamation phase.

(5)The period since middle and later Yanshan movement at that time the edge of continent was segregated and split.

During middle and later Yanshan period, the segregation and splitting in East China which was represented as Tai-Lu fault were occurred. New segregated and split plate started to generated. The group of rift basins in Mesozoic and Cenozoic appeared. And Non-marine Mesozoic and Cenozoic sediments were deposited, they constructed the newest sediments covers in the area. This period was called as continental edge segregation and split phase.

7. The Evaluation for oil and gas resources

There is a very close relationship between the evaluation for oil and gas resources and the geology event in deep. Because of the hydrocarbon source rock is a part of basin, and basin is a part of lithosphere. The evolution of basin and source rock were controlled by the evolution of lithosphere, while some factors after evolution, such as the state of hydrocarbon source rock, the condition of preservation, the ability of hydrocarbon reconstruction, and the assemblage of pool generation and so on, all these, were the determinate factors which controlled amount and quality of oil and gas fields. In particular, the oil and gas fields contributed by non-organic material were just the result by direct action of geologic events in deep.

In this area, most organic hydrocarbon source rocks were in marine faces, up to now, the gas was the main result of evolution. Hydrocarbon was generated in multiple periods, pool was composed in multiple periods too, but the later one was in the most favorable state. In our project, the maps of relic thickness of main marine face source rock in Mesozoic and Cenozoic were completed, it makes breakthrough at the evaluation of oil and gas prospect. According to relic thickness map and the difference in preservation condition. Five type of hydrocarbon exploration domains associated with marine face source rocks were pointed out. They are 1. the area covered by non-marine face Mesozoic and Cenozoic state, 2. the area covered by nappe (metamorphic rock), 3. the area covered by volcanic rock, 4. the coalificational gas and 5. the close sand petroliferous strata in Longtai coal measures (the last two parts were preserved in coal measures strata). These five domains were added to non-marine face middle-small basins in Mesozoic and Cenozoic Era, shallow gas and inorganic hydrocarbon, these eight domains show clearly the areas and aspects for the oil and gas exploration in the future.[4]

It is worth pointing out that in this project the genesis conditions of inorganic hydrocarbon pool were discussed systematically, and two kinds of patterns in pool genesis were suggested. The keys for genesis conditions of inorganic pool are (1) the deep and huge fracture in lithosphere should exist, the fracture acted as a channel, let the inorganic hydrocarbon along it move from deep to crust surface.(2) during the inorganic hydrocarbon raised up to surface no oxidation happened.

According to these two conditions, two possible patterns for pool genesis were firstly purposed.

(1) In the rift systems with reducing environment and thick sediments(lower oxygen fugacity) or down-faulted basin(with deep and huge fractures as channels), the inorganic hydrocarbon came from the deep without oxidation, entered the various traps in the basin, then the pools were formed, frequently they mixed with organic hydrocarbon to from pool.

(2) Inorganic hydrocarbon source was accumulated in the horizontal fracture zones of uplifting area. The horizontal fracture zones raised along with the area uplifted, the undamaged relic of horizontal fracture zones (they were the ideal fracture traps) and the inorganic oil and gas in them might be preserved and migrated to surface (i.e. the depth that human beings can perform exploration and production with morden technology),to become inorganic oil and gas pools, available for people.

In China, the Mesozoic and Cenozoic rift belts and down-faulted basins as well as the associated uplift areas all are the favorable potential areas for finding inorganic hydrocarbon pools. The Lower Yangtze area and its adjacent region is one of such area.

In the eight domains for oil and gas exploration in this area, the most favorable domains include the marine face covered by non-marine Mesozoic and Cenozoic rocks, shallow gas and inorganic hydrocarbon. From now on our main target should put on the finding of large and middle gas field, to perform gas survey widely. In the survey, geochemistry prospecting particular in hydrocarbon detection should go ahead, after the geochemical anomalies were discovered , then geophysical prospecting and drilling are followed.

REFERENCES

1. Chen Husheng: Comprehensive geophysical survey of HQ-13 Line in the Lower Yangtze Region and its geological significance, OIL & GAS Geology, 1988(3), 214-216.

2. Chen Husheng: Descend-raised undation and its significance in looking for oil and gas,Journal of Xian College of Geology, 1984(1), 98-100

3. Chen Husheng: some problems concerning Geotectology, Geophysical Prospecting for Petroleum, 1986(2),5-6

4. Chen Husheng: Petroleum Prospecting in the Areas of the East China During the Time of 1990s, Experimental Petroleum Geology 1992(1), 4-6

5. Wang Hongzhen: A reprospect of the study of global tectonics, Earth Science Frontiers, Vol.2 No.1-2, 1995

6. Xiao Qinghua Li Xiaobo et al. : Frontiers on orogenic belt researches, Earth Science Frontiers, Vol.2 No.1-2, 1995

7. Kenl C. Condie : Plate tectonic & Crustal Evolution, Scientific press, 1986

8. Edited by Chinese geology society, Modern World Geoscience Trend-Chinese scientists talk on the 28th International Geology Congress, Geology Press, 1989

9. Edited by China Geoscience Institute: The Basin Problems and Mathedlogy in Lithosphere Research, Metallurgical Industry Press, 1990

Proc 30* Int'l. Geol. Congr., Vol. 4, pp. 119-125
Hong Dawei (Ed)
© VSP 1997

THE LATEST DEVELOPMENT IN CONTINENTAL SCIENTIFIC DRILLING IN CHINA(CSDC)

LIU GUANGZHI

Member of Chinese Academy of Engineering.   MGMR, Xisi, Beijing 100081, China

Since the 90's, an encouraging achievement has been obtained in the preparation of the Continental Scientific Drilling in China(CSDC). A seminar on the key geological problems in the research of China deep geology was convened in March of 1991. The priority Project of Continental Scientific Drilling in China(CSDC) was implemented (July,1991).The concerned report was submitted (Nov.1994) and then was appraised.It was considered that the conditions were ripe for conducting continental scientific drilling in China. We strove to get this project listed into the country's major scientific Engineering projects in China's Ninth Five-Year Plan. The State Council issued the National Program for Medium and Long Term Scientific and Technological Development, indicating that the technical preparation should be conducted before the year of 2000 for geological scientific (ultra-deep) drilling engineering, and then geological scientific (ultra-deep) drilling project would be implemented before 2020 (Feb. 1992). The first (April, 1992) and second (May, 1993) seminar on China Continental Scientific Drilling were held . The site selection for China's scientific drilling was discussed. Twelve candidate sites were discussed, then summarized into four, and finally Dabie-Jiaonan was selected as the first target area. China National Geological Super-deep Drilling Laboratory was set up in China University of Geosciences(Beijing), and the equipment installed (May, 1993). A delegation was sent to attend the International Conference on Continental Scientific Drilling in Potsdam, Germany in August, 1993 and also participate the preparatory meeting of ICDP (Sep. 1993).Mr. Xiao Xuchang was recommended as the member from China by the Ministry of Geology and Mineral Resources (Jan. 1994).Mr. Xiao Xuchang and Mr. Min Zhi attended the ICDP meeting in Stanford University (Dec. 1995). The implement of China's First Continental Scientific Borehole Drilling and Its Scientific Research was formally submitted and approved to be the third item in China's Mega-scientific engineering projects during the period of Ninth Five-Year Plan after the discussion and vote by the senior experts organized by China's National Science Commission (Feb. 1992).The Ministry of Finance agreed to pay the fee for ICDP (Jul. 1995).The 36th Xiangshan Scientific Conference under the subject of Continental Dynamics and Continental Scientific Drilling was convened. On this conference, a common understanding was obtained about the implementation of Continental Scientific Drilling in China (May, 1995). The Third Seminar on China's First Continental Scientific

Borehole Drilling was convened to finally determine Dabie-Jiaonan as the target area. Experts were organized to visit and investigate the site (Jan.1996). The Initial Technical Research and Development Program for China's First Scientific Borehole Core Drilling was put forward so as to accelerate the research and development in drilling techniques. Following is the brief introduction on the R & D of the drilling technology for the first scientific borehole about the National Geological Super-deep Drilling Laboratory.

SCHEME ON THE R & D OF THE DRILLING TECHNOLOGY OF THE FIRST SCIENTIFIC BOREHOLE CORE DRILLING IN CHINA

The symposium on site location of Dabie-Jiaonan was held in Anhui province in January, 1996. The work which running parallel with the site location is the R & D in drilling, ignoring this will delay the whole progress of the construction of the first pilot hole.

Some important strategy and concept which should be considered

1. China is a developing country, so most of the equipment, tools and technology should be developed largely through our own efforts. The potentiality of our drilling experiences accumulated for more than 40 years should be brought into full play as many as possible. Some key products should be imported from foreign countries.

2. In drilling the scientific pilot hole(with normal depth 4000-5000m.) the following measures strategy must be considered: (1) Slim-hole diamond core drilling should be employed in all sections to reduce the volume of rock crushing and to increase the penetration rate, except in hole-opening section. (2)In order to reduce the tripping time by a big margin, reduce the construction time effectively and improve the quality of cores, wire-line core drilling system must be used. Bit change without hoisting and lowering drill rod or with less hoisting and lowering drill rod can reduce labor intensity and save one third of the drilling cost. (3)PDM and hydro-hammer driving wire-line coring tools should be developed and employed by taking advantage of the technological superiority of our country in slim-hole downhole motor(DHM) along with wire-line coring system to accomplish slim-hole drillstring non- or less rotation drilling which can save power, reduce wear of drillstring and casing, effectively prevent the bore hole from deflection. (4)physical-chemical methods are to be mainly used to stabilize bore hole wall. Casing tubes are to be run only in upper section, open hole drilling is employed as long as possible in lower section of crystalline rock. Small clearance casing plan should be adopted when rock formation is unstable in troublesome zone. (5)The function of multi-information carrier of drilling fluid should be brought into full play to transfer a large amount of geological information (deep fluids and cuttings)(fig. 2).

Preliminary R & D project

Principles of Establishing the Project: 1. Taking crystalline rock (with uniaxial compressive strength all above 1000-1500kgf/cm) as the main object of drilling; 2. Taking 350-200°C as bottom hole temperature in design; 3. Taking pilot hole as the first, on this base, deep hole and super deep hole can be designed.

Surface Equipment: 1. Top drive drillrig for wireline coring and its guide; 2. Automatic draw-works; 3. Full automatic drill rod handler; 4. Full automatic drill rod break-out tool; 5. Drill rod fatigue and broken detector; 6. Integrated control console for drilling technology(monitoring, collection, treatment, optimization, feed-back system)

Research on Theories of Drilling Technology: 1. Rock fragmentation mechanism and down hole process during drilling under high temperature and high pressure;classification of drillability of crystalline rocks; 2. Establishing high pressure or high temperature test facilities; 3. New materials of light weight and high strength tubings; 4. Drill rod fracture mechanics and monitoring system; 5. Borehole structure, casing program; 6. Mechanism and rule of bit wear; 7. Establishing of data bank of opportunity wells;

System in the Borehole(Table 1): 1. Long service life diamond bits and reamers; 2. Developing super-hard cutters and coring bits; 3. Wireline coring system driven by down hole motors (DHM)(hydro-hammer, PDM): (1) Hydro-hammer driving wireline coring system; (2) Positive displacement motor driving wireline coring system; (3) "Three in one" (WL+PDM+RTB) Coring System; 4. Sidewall coring tools: (1) Hydraulic or scraping; (2) Horizontal drilling of down hole electric motor; 5. High pressure and high temperature liquid and gas sampler; 6. Down hole small-bore MWD instruments; 7. Small-bore Vertical Drilling System(VDS); 8. High temperature stability, high lubricate, anticorrosion drilling fluid and its additives; 9. High temperature cement and cementation technology; 10. Formation tester flowmeter, etc.; 11. Deep hole core orientation(high accuracy, thermal stability and high dependability); 12. Titanium bearing aluminum alloy drillrod (thermal stability, fatigue resistance, abrasive resistance and embattlement resistance)

Table 1   Temperature limitation for current drilling tools, instruments and materials in borehole

Item	Journal bearing rock bit	Diamond bit	PDM	percussion tool Hydro-hammer	MWD	Cement	Drilling fluid	Casing drill rod
Temperatre limitation	300 ℃	400 ℃	147 ℃	295 ℃	145 ℃	300 ℃	180 ℃	400 ℃
Weak points	Compressive resistance of material	-	Bearing stator rubber seal only 100-120m service life; compressive resistance	Percussion element seal	Short servicelife, seal and compressive resistance element	Rheological feature, fracture and foreign additive	Rheological and stability feature destroyed	-

Note: 5500 meters logging cable is to be imported.

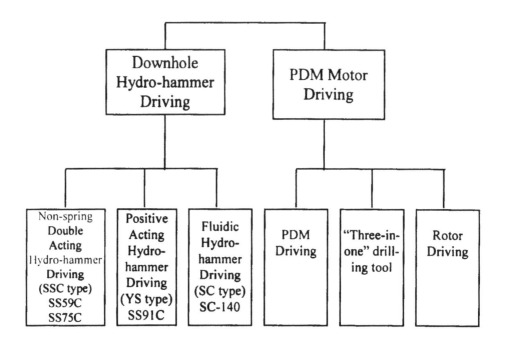

Fig. 1   Wireline coring system driven by downhole motor (DHM)

Drilling Fluid and Additives: The temperature at the depth of 4000-5000m. can reach 150 ℃ according to the normal geothermal gradient. The high temperature will increase the dispersity of clay in drilling fluid, make drilling fluid thickened and increase oxidation and breaking of molecular bond of micro-molecule chemical additives, all of these will make the additive failure, sharply increase the fluid water loss and make its stability, flowability and tixotropy destroyed, so as to influence the wall protection.

Table 2   Additives with temperature resistance of about 200 ℃ made in China

Additive	Temperature resistance
Hydrolyzed polyacylonitrile	200-300 ℃
Ferrochrome lignosulphonate	130 ℃
Humic acid and its derivatives	
1. Sodium humic acid (NaC)	200-230 ℃
2. Chrome-sodium humic acid	230 ℃
3. Chrome-sulfonated-methyl-humic acid(SMC)	220 ℃
Sulfonated methyl gallic tannin (SMT)	180-200 ℃
Carboxymethyl cellulose (CMC)	140-180 ℃
Na-HPAN	>200 ℃
Ca-HPAN	>200 ℃
Geothermal 93	260 ℃

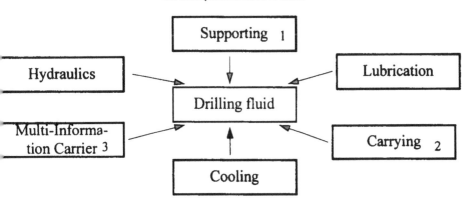

Fig. 2 The multi-information carrier function of drilling fluid

Note: 1. Indicating well kick, blowout, collapse prevention and borehole diameter reduction; 2. Indicating carrying cuttings and chips; 3. Carrying the deep formation fluid (oil, gas and water, etc.) to the surface, such as H_2, O_2, CO_2, CH_4, He, H_2S, SO_4 as well as Na, K, Ca and Mg micro particles.

Information Acquisition: Coring, Sampling, Borehole Logging and Fluid Test

Conscientious sampling of typical samples and acquiring more data are the primary request for the success of scientific drilling. Data can be gathered from inside hole and at the surface. Inside hole data gathering includes coring, sand sampling, logging, drilling and hydraulic testing, inside hole hegphysical experiment(between hole bottom and ground surface). Data gathering at ground surface includes solid and liquid samples acquired from the hole top and drilling fluid monitoring instruments, including the first and primary geological description, chemical and physical analyses on these samples. Logging project is ambitious and time consuming. The cost for drilling pilot hole is equal to the cost of borehole logging. Logging program should be executed at one section after the other. In order to prevent the danger(fishing operation) due to the missing of data, institutions of the universities will undertake monitoring and experimentation for a long period of time after drilling operation is completed.

THE NATIONAL LABORATORY ON SCIENTIFIC DRILLING

A. outline

The National Laboratory on Scientific Drilling was set-up in May 1993. It is an open research laboratory

.

B. The study Orientation

1. The pre-study of scientific drilling and super-deep drilling technology; 2. The study of computer application techniques in drilling field.

C. Form of The Laboratory

The major equipment include: computer-control drilling test system, some of personal computers and their peripheral, and a HP9000 workstation.

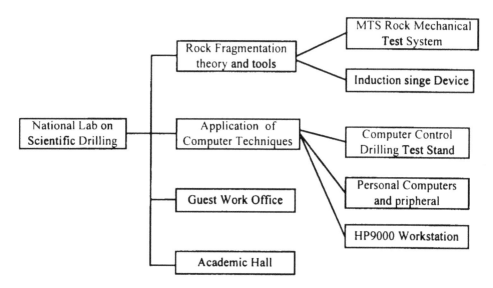

D. Achievements

1. Automatic computer-control drilling test system; 2. Computer analysis system on drilling technology; 3. Research of diamond drilling theory and optimization drilling technique; 4. Study on new AC driving core drill system; 5. Study on situation of drilling stem and bit weight control; 6. Retrievable bit techniques(corporation with Russia); 7. Rock melting techniques(corporation with Russia)
8. Study of high temperature drilling fluids;

Imitating HTHP experimental equipment

Experimental Functions: 1. Imitating research on drilling technology; 2. Rock mechanics research under HTHP; 3. Research on fracturing and permeability theory under HTHP; 4. Strength test of pipe under HTHP; 5. Research on borehole wall stabilization; 6. Research on formation and evolution of rock and mineral; 7. Research on high temperature drilling fluid; 8. Research on control in downhole process; 9. Experiment on the HTHP resistance of geophysics and MWD

HT-HP Test Device for Geosciences

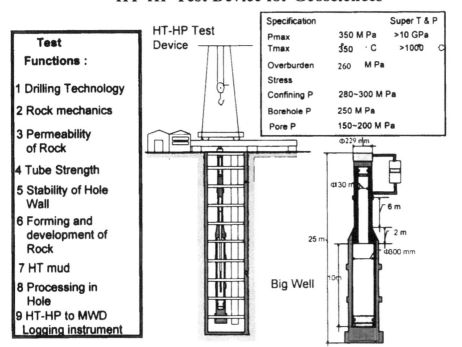

Test Functions :

1 Drilling Technology

2 Rock mechanics

3 Permeability of Rock

4 Tube Strength

5 Stability of Hole Wall

6 Forming and development of Rock

7 HT mud

8 Processing in Hole

9 HT-HP to MWD Logging instrument

HT-HP Test Device

Big Well

Specification		Super T & P
Pmax	350 M Pa	>10 GPa
Tmax	350 · C	>1000 C
Overburden Stress	260 M Pa	
Confining P	280~300 M Pa	
Borehole P	250 M Pa	
Pore P	150~200 M Pa	

Proc. 30ᵗʰ Int'l. Geol. Congr., Vol. 4, pp. 127-133
Hong Dawei (Ed)
© VSP 1997

Technical Concept for the Drilling of a 5000 m Deep Scientific Core-hole

Zhang Wei

Chinese Academy of Geoexploration, 31 Xueyuan Road, Beijing 100083, China

Abstract

Special drilling technologies are needed for the drilling of a 5000m deep scientific core-hole. The general criteria for drawing up technical concept is to optimize the economics of drilling operation under the prerequisite of meeting the requirements of geoscientific research and of being technical feasible. The end diameter of the borehole should not be less than 6-in to meet the requirement of logging operation. According to the technical level and current status of the drilling techniques in the world, a hybridized drilling technology(petroleum drilling/mining drilling) should be adopted for the drilling under special and hostile drilling conditions. This technology is characterized by installing a high speed top-drive head and a diamond wireline coring system onto a petroleum rotary rig, which has been proven in praxes to be the most cost-effective method for continuous coring in deep and hard formation.

Keywords: Technical concept, Scientific drilling, Deep core-hole

Introduction

China has become a member of ICDP and is now making preparation for implementing a scientific deep drilling program in Jiaonan-Dabie area of eastern China. A 5000m deep scientific borehole is planed to be drilled in crystalline rock formation, to investigate the world largest ultra high pressure metamorphic belts there and reconstruct the exhumation history of the ultra high pressure metamorphic rocks

1. The General Criteria for Drawing up the Technical Plan

The drilling of a 5000 m deep core-hole in crystalline rock formation means a strong challenge to the drilling technology. The general criteria for drawing up the technical plan is to optimize the economics of drilling operation under the prerequisite of meeting the requirements of geoscientific research and of being technical feasible. The following three principles should be followed to reach these goals:

-- Using as much of the existing drilling technologies as possible and conducting R&D projects if necessary indeed;

-- Mainly using drilling equipments , tools and instruments available on domestic market whereas considering importing a few key parts that are not available at present in China ,

-- Combining the technical capabilities of geo-drilling and petroleum drilling to resolve the technical problems that can be encountered while conducting continuous coring in a deep borehole .

2. The Basic Borehole Parameters Needed for Drawing up Technical Plan

2.1 Borehole Depth

According to the results of preliminary geological and geophysical investigations, the maximum borehole depth of the first borehole of Continental Scientific Drilling of China (CSDC) is within 5000 m.

2.2 Borehole Diameter

Borehole diameter is a very important parameter in the technical plan of drilling and will affect, to a very large extent, the technical concepts of drilling and logging, the scientific yield and the economics of the entire program, so it should only be determined after all the factors have been carefully considered.

If we only consider lowering the operation cost, the borehole diameter should be as small as possible. But the borehole diameter should be large enough to enable the successful implementation of a logging program, because logging is one of the most important means of obtaining geoscientific information.

According to the currant status of logging technique, for the logging in a 5000 m deep borehole, the end diameter of the borehole should not be less than 6-in(152mm), if a lot of logging methods are to be used in the same borehole.

2.3 Percentage of Core Drilling

To meet the requirements of geoscientific research, continuous coring is considered, namely the percentage of core drilling is 100%.

2.4 Formation Conditions

Crystalline rocks with high hardness, strength and abrasiveness, such as gneiss, quartzite, eclogite and so on, are expected to be encountered in Dabie Mountain area.

2.5 Coring Method

It has been proven by the praxes both in China and abroad that diamond wireline coring is the most cost-effective method for the continuous coring in a deep and smaller diameter borehole existing in hard rock formation . This coring method is believed to be the most ideal one for the future Dabie Scientific Drilling Project.

3. The Proposed Casing Program

A $14^3/_4$-in(348mm) tricone bit is to be used to spud in. After the loose and unstable overburden formation has been passed , a $11^3/_4$-in (298mm) conductor casing will be set and cemented, and then a 7-in retrievable casing hanged onto the casing head. The core drilling will begin with a 6-in(152mm) diamond core bit and a $5^1/_2$-in (140mm) wireline coring drill string. It is ideal that we can reach the planed final depth without any reaming and casing-setting needed(seeing Fig. 1a) . If the unstable formation is encountered, we can at first remove the 7-in retrievable casing string, and then enlarge the borehole. The enlarged borehole section should pass the unstable formation, in order that the casing string can seat on a stable base. A larger casing string will be then installed and cemented . After that the 7-in casing string will be once again hanged on to the casing head and the drilling with 6-in diamond core bit continued. In this casing program, we have two casing strings as reserve, a $8^5/_8$-in string and a 7-in string respectively. (seeing Fig. 1b and Fig. 1c). In fact this casing program is based on the ideal of " the Advanced Open Borehole Method", a Russian approach [1]. This approach has been proven, through large amount of activities of Russian Continental Scientific Drilling, to be the best suitable method for drilling under the uncertain underground conditions.

4. The Technical Concept for the Drilling Operation

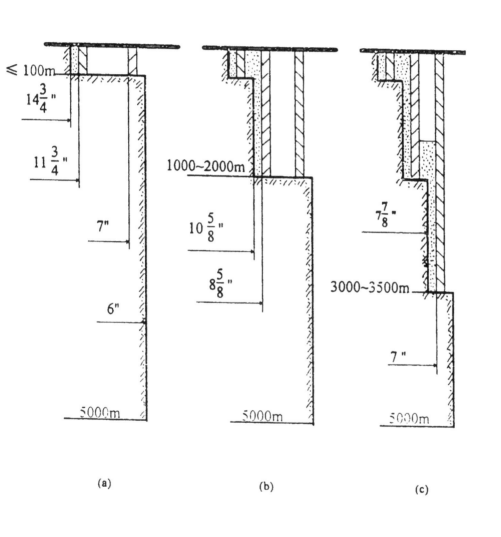

(a) (b) (c)

Figure 1. Casing Program of the 5000m Deep Scientific Core-hole

4.1 Drill Rig

The total 5000 m of $5^1/_2$-in wireline drill string weighs about 135t. After considering the safety factor of 1.8, the hook last of the drill rig should not be lower than 240t. The world's largest core drill for geological exploration has a maximum hook last of 120t, which is far from meeting the requirements of drilling operation of a 5000m deep scientific borehole.

The rotary drill rigs for petroleum drilling are powerful enough, but the rotational speed of their rotary tables is too low to meet the requirement of diamond core drilling.

By combining the technical advantages of geo-drilling and petroleum drilling, namely by incorporating a high speed electric(or hydraulic) top-drive head and a diamond wireline coring system into a medium heavy petroleum rotary rig, we get a " hybrid approach" of drill rig, which has been proven, with its reliability and cost-effectiveness, respectively in Germany [2], the USA [3] and UK.

4.2 Wire-line Coring Drill String

Drill String is the key element of the whole drilling system, especially for the wireline coring in a borehole as deep as 5000m . For coring in 6-in borehole, a $5^1/_2$-in(140mm) steel wireline coring drill string is proposed to be used.

4.3 Downhole Coring Systems

To improve drilling performance, combined downhole coring systems should be used. These systems , such as retractable bit system, retrievable hydro-hammer system(Fig.2), downhole-motor/retractable bit combination system(Fig.3)[4] and so on, incorporate some advanced downhole drilling mechanisms into the basic wireline coring system. They have very bright prospective for coring in deep and hard rock formation. But further R&D work should be done to increase their reliability and effectiveness for working under hostile borehole conditions.

4.4 Core Bit

Mainly impregnated diamond core bits are to be used. For coring in deep and hard formation, the improvement of the diamond core bit should be concentrated on increasing rate of penetration under the prerequisite of having enough long bit life. Special concern should be given to the gauge protection of the bit against the rapid wear because of the hard and abrasive rock formations.

4.5 Drilling Fluid System

The drilling fluid system used for drilling a 5000 m deep scientific borehole should meet simultaneously different requirements. A low solid content or solid free drilling fluid system with thermostability as high as 150 ℃ and having low flow and friction resistance, strong cutting carry capacity and having no negative effects on geochemical analysis is expected to be used.

5 Conclusions

The Drilling of a 5000 m deep core-hole in crystalline rock formation in China is technically feasible. Except for a few key parts of drilling systems that are not available at present on Chinese market and should be introduced from foreign countries, the most parts of drilling systems including drill rig, most of drilling tools, materials and instruments can be provided by domestic manufacturers, suppliers and research institutions. But because of the extreme drilling conditions of such project, a lot of R&D efforts should be made to improve the reliability and effectiveness of the existing drilling systems.

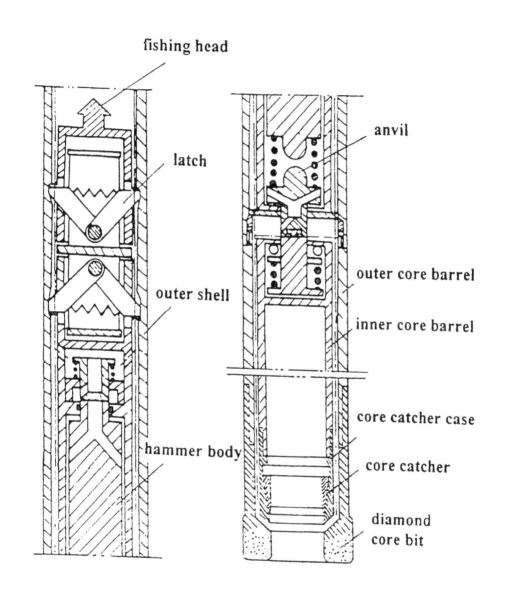

Figure 2. Retrievable Hydro-hammer Core Drilling System

132

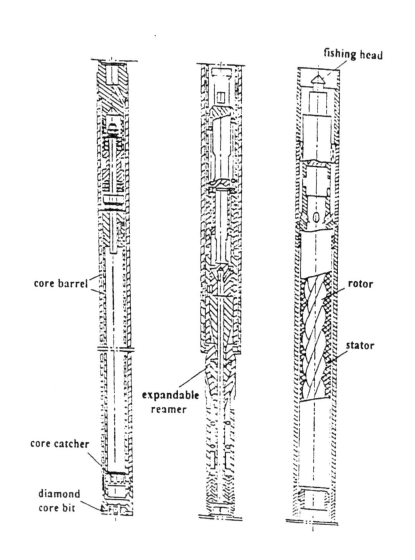

Figure 3. Downhole Motor/Retractable Bit Core Drilling System

References

1. E. A. Kozlovsky. The Superdeep Well of the Kola Peninsula, Springer Verlag, Berlin, Heidelberg, New York, 1987, Chapter 3 " drilling".
2. H.Rischmueller and C.Chur. Kontinentales Tiefbohrprogram der Bundesrepublik Deutschland-- Technische Konzeption und Stand der Planung, Erdoel , Erdgas und Kohle. 1987, No. 1.
3. B. Wagner. Successful Hybridizition of Drilling Technology for Reliable Continuous Coring Systems. Proceedings of the 4th International Symposium on the Observation of the Continental Crust through Drilling, 1988
4. Liu Guangzhi. A Preliminary Plan for the Continental Scientific Drilling in China. Scientific Drilling, 1992, No. 3.

Proc. 30 *Int'l. Geol. Congr.*, Vol. 4, pp. 135-141
Hong Dawei (Ed)

THE GRAVITY FIELD AT THE CONTINENTAL MARGIN OF SOUTH AMERICA (20° to 29° S)

H.-J. Götze and A. Kirchner

Institut für Geologie, Geophysik und Geoinformatik, FU Berlin, Haus N, Malteserstraße
74-100, D-12249 Berlin, Germany

INTRODUCTION

From 1993 to 1996 the international MIGRA group with participants from Chile, Argentina
and Germany has surveyed some 3.000 new gravity observations in an Andean Geotraverse
covering N-Chile and NW-Argentina between 64°- 71° W and 20° - 29° S. MIGRA is a
Spanish acronym for "Mediciones Internacionales de GRavedad en los Andes" [1]. Including reprocessed older data of Freie Universität Berlin, South American Universities, Oil
and Mining Industry, there is a data base of more than 15.000 gravity values available, which
can be used together with other geophysical and geological information for an
interdisciplinary interpretation of the structure and evolution of the Central Andes [2].

In summer 1995 MIGRA took part in the "CINCA" offshore experiment of the german
research vessel "Sonne" between the latitudes 20° S to 24° S. The offshore gravity data is
connected with the on land survey to draw a complete gravity/geoid picture of this ocean-
continent transition. Gravity data at sea were collected along continous survey lines (some
8.000 km). The average density of observational sites amounts of about 15 observations per
km. Gravity survey of cruise SO - 104 was tied to the Chilean National Gravity network of
the "Departamento Geofisico, Universidad de Chile, Santiago" at reference stations in
Valparaiso, Antofagasta and Iquique. The drift of the KSS31/32 onboard gravity sensor was
very low and results in -0.048 mGal/day or - 1.44 mGal/month, respectively. The overall
drift of the gravity meter was determined to be 0.47 mGal per 92 days.

ON- AND OFFSHORE GRAVITY DATA

The investigated region covers a 900 km x 1.000 km area in the central part of the Andean
orogenic system. The young Andean orogen between 20° - 29° S comprises different
structures which have evolved on a Precambrian-Paleozoic basement (e.g. [3] and [4]).
Several age determinations of granitic stocks and high grade metamorphic rocks in N-Chile
and NW-Argentina suggest that an extensive NNW-SSE striking belt of Early Paleozoic/Pre-
cambrian basement dominates the crustal structure. This belt of ancient rocks was also
described as the border of the "Faja Eruptiva Occidental". Two of our gravity surveys
obtained structural information with new stations covering the northern and southern edges
of this belt, near Calama (Chile) and in the Southern Argentinean Puna respectively. The
investigated area is characterized by its enormous topography and remoteness, by its aridity,
low population density and limited infrastructure. Other difficulties limiting our field work
were the lack of topographic maps and geodetic networks in some regions. The spacing of
stations amounts to approximately 5 km along all passable tracks aside from some local
areas with a higher station density . To complete this data base we included gravity observa-

"Eastern" and "Western Cordillera", the gravity coverage for the region is fairly uniform. All measurements are tied to the IGSN71 gravity datum at base stations 40334K in Oran/Argentina, 40400K in Iquique/Chile and 40365L in Tucumán/Argentina.

The large size of the area and the severe logistical problems did not always allow us to determine the drift of the gravity meters by repeating the measurements at each station. However, even when we used bad tracks, the drift of the LaCoste & Romberg instruments (model G) rarely exceeded 0.1 mGal per day. Only about 35% of the gravity sites could be tied directly to benchmarks, such as levelling lines or trigonometric heights, so we used altimeters for height determinations. To improve the quality of our barometric measurements, we calculated time-dependent drift corrections as it is usually done for gravity measurements, using as many benchmarks and repeated measurements as possible. Moreover, the profiles of several days were tied together in order to eliminate systematic errors. The scales of the barometers have been calibrated on levelling lines with an altitude difference of about 2.000 m. Error estimations showed that even in the worst case the accuracy was better than 20 m, giving an error in the Bouguer anomaly of about 4 mGal, which is less than 1% of the overall magnitude of more than 450 mGal.

For the terrain correction (up to 167 km around all stations), a method including calculations of the earth´s curvature developed for gravity investigations in the Alps was used, after adapting it to the special situation in the Central Andes. Reduction density was 2.67 g/cm^3 and the digital terrain model by [5] was used to calculate true 3D terrain corrections. For different morphological units we obtained the following typical values of topographic reduction: Longitudinal Valley and Chaco region: 0.5 - 1 mGal, Coastal, Pre- and Western Cordillera, Altiplano/Puna and Subandean Belt: 1 - 10 mGal and steep coast and Eastern Cordillera: 10 - 25 mGal.

GRAVITY ANOMALIES

Onshore the Bouguer anomaly drops down to a regional minimum of about - 450 mGal in the area of the recent volcanic arc, related to crustal thickening by isostatic compensation. The effect of isostatic compensation of topography was calculated assuming the model of Vening-Meinesz with the following parameters: density contrast of the earth´s mantle and crust $\Delta\rho = 0.35$ g/cm^3, normal crustal thickness: 35 km and a flexural rigidity of 10 Nm. The gravity effect of the isostatic compensation root was eliminated from the Bouguer gravity and the resulting anomaly serves as a residual field (Fig. 1). The most interesting features of this field are: (1) Positive values in the area of the forearc with isolated complexes parallel to the coastline. They are regionally caused by the presence of the dense subducting plate (gravity effect of about 50 mGal; density contrast: 0.05 g/cm^3) and locally by uplifted jurassic batholiths intruded into the "Formación La Negra". (2) The NNW-SSE striking positive anomaly from Calama (CAL) by the Salar de Atacama to southern Puna. We explain this gravity maximum by a highly metamorphic and high-density Paleozoic/Precambrian structure, the "Faja Eruptiva Occidental", which is oblique to the N-S orientation of the recent volcanic belt. (3) Local minima along the recent volcanic arc point to reservoirs of partly molten material at depths of 15 - 20 km. (4) Minima following a line from Ollagüe (OLL) to Calama (CAL) along 69° W, are caused by the Eocene volcanic arc with low-density volcanic material in the upper crust. (5) Alternating gravity highs and lows in the backarc region east of 67° W are observed in wide areas of the Argentine Puna and the Eastern Cordillera with a general NE-SW trend. The minima point to the position of mesozoic basins which are located in the Argentine Puna and extend northward to the territory of Bolivia. Gravity highs correlate with outcrops of Precambrian/Paleozoic basement in the

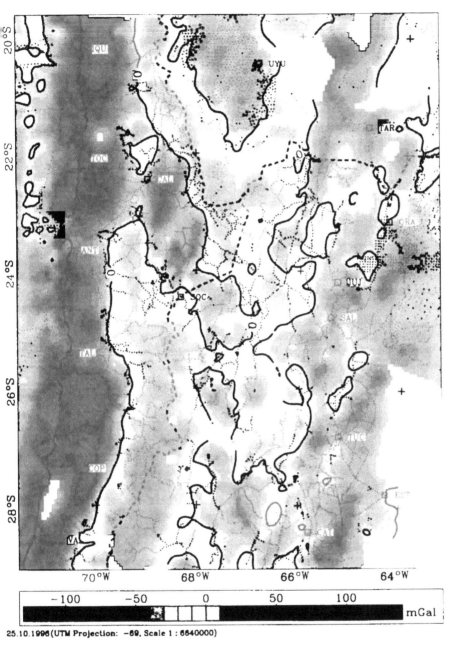

Fig. 1: Residual gravity field of the Central Andes. Color shades contour gravity by 10 mGal intervals. Gravity stations are shown and the 0–mGal contour line.

Puna and Eastern Cordillera.

CRUSTAL DENSITY MODELING

3D forward modelling of both gravity and gravity potential (geoid) can explain that most of the observed geoid anomaly (undulations) up to 50 m is caused by Andean topography and isostatic roots. However, also density inhomogeneities in the downgoing slab and in the asthenospheric wedge contribute to undulations of the Earth's geoid in the Central Andes. All density modelling has been proven by the results of refraction seismics (e.g. [6]).

A large scaled 3D density model was constructed to investigate the regional structure and density distribution of the Andean lithosphere. The modelled area reaches from 12° to 35° S and from 57° to 79°W comprising large parts of both the Nazca Plate in the west and the Brasilian Shield east of the Andean orogen as a reference lithosphere. The principal parts of the model are the downgoing Nazca Plate, Andean crust, continental lithosphere with Brasilian shield crust and an astenospheric wedge.

The model resembles the new results of refraction seismics based on two WE and two NS trending ray tracing models at 24° and 23.5° and 68° and 69° W respectively as described above. Additionally, a balanced cross section at 21° S according to earlier refraction seismics [7] was included. As a preliminary study the velocities of seismic models were directly converted into densities by using e.g. the Ludwig - Nafe & Drake relationship or similar velocity-density relationships. Initial model geometry was slightly modified by the use of interactive computer graphics to verify regional trends implied by the Bouguer gravity field and the topography-reduced geoid. Fig. 2 shows the major crustal parts of the density model at 23.5° S.

In the forearc a 10 km thick upper crust with a density of 2.73 - 2.75 g/cm^3 tops a 20 km thick high-density lower crust material (3.00-3.03 g/cm^3), followed by parts with reduced density (2.93-2.90 g/cm^3) which are interpreted as serpentinized Pre-Andean mantle material. In the Precordillera the upper crust thickens to about 20 km with its density decreasing to 2.69 g/cm^3. It can be followed to the recent arc with the same thickness and again decreasing density (2.67 g/cm^3). Towards the back arc the upper crust base rises to 15 km, its density increasing to Precordilleran values. The Coastal Cordilleran high-density material can also be pursued in WE direction with decreasing densities down to 2.86 g/cm^3 and its base reaching 40 km as it passes beneath the recent arc. There is a low-density zone of 2.70 g/cm^3 underneath the volcanic chain between 20 and 35 km depth resembling the LVZ detected by refraction seismics.

Lower crustal bodies are modelled with a density of 2.98-3.00 g/cm^3 below 40 km placing the "gravimetric" moho to a depth of approximately 60 km according to 2D ray tracing models. A seismically implied second low-density zone is modelled between 44 and 50 km with a density of 2.92 g/cm^3. The Brasilian shield crust consists of a 4-6 km thick sedimentary cover (2.50 g/cm^3) followed by a 2.83 g/cm^3 mid crustal body and a lower crust down to 40 km with a thickness of 10~km and a density of 3.10 g/cm^3. The shield crust lies on top of the continental lithosphere modelled with a density of 3.37 g/cm^3.

The shape of the subducting Nazca Plate is inferred from results of [8] and earthquake hypocenters of the PISCO 94 seismic catalogue (Asch, Rudloff, Graeber, pers. comm.). For definition of the descending angle in the forearc there are moho observations by refraction seismics between 21° and 24° S. The transition of a 2.90 g/cm^3 oceanic crust to a 3.05 g/cm^3

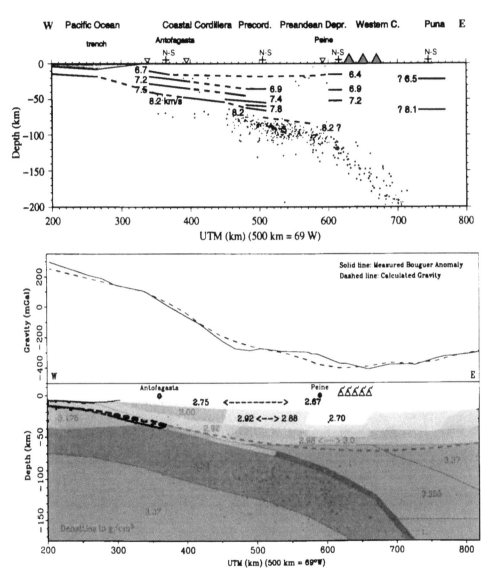

Fig. 2: A cross section of the 3D density model along latitude 23.5 S. The image contains the density domains and shows the constraints that were used from seismics research and seismology (e.g. Schmitz 1994; Wigger et al. 1994). The assymetry of the gravity field reflects the different crustal behaviour in the forearc with a high density distribution and in the backarc with lower density. For comparison the isostatic crust–mantle boundary of a Vening–Meinsz model (T=35 km, drho=0.35 g/ccm, D=1e23 Nm).

eclogite layer on top of the downgoing slab, which starts at about 50 km and is completed at 70 km depth, is simulated by a model body with a density of 3.15 g/cm^3 of oceanic crustal thickness within this depth range. In the model we present here the subducting Nazca Plate is modelled down to 670 km with a mean density contrast to the surrounding asthenosphere of 0.02 g/cm^3.

The model also contains an asthenospheric wedge with a density of 3.36 g/cm^3 filling the space between the slab and the shield lithosphere. According to studies on Andean magmatism and P-wave attenuation (e.g. [9]) it reaches the crustal root beneath the recent arc with an alongstrike change of geometry. The updated gravity data base will play an important role in both local investigations of applied geophysics and regional interdisciplinary interpretations of pure geophysics and geodesy [10]. Andean gravity field seems to be a sensitive indicator which is linked to many processes contributing to the tectonic framework of the Nazca plate subduction zone.

Acknowledgement: The support with maps and geodetic information by the following institutions is gratefully acknowledged: Institutos Geográficos Militares in Santiago, Buenos Aires and La Paz, Yacimientos Petrolíferos Fiscales of Argentina and Bolivia, Empresa Nacional de Petróleo, and CODELCO both from Chile. This paper presents results of the gravity research group D3 of the SFB 267 which is funded by the Deutsche Forschungsgemeinschaft (DFG) and Freie Universität Berlin and contributes to IGCP 345. MIGRA data sets are available via FTP for non commercial applications of universities and governmental agencies. Please contact H.-J. Götze at Freie Universität Berlin (Germany) under "hajo@zedat.fu-berlin.de" or refer to the Website of the research group for further information: http://userpage.zedat.fu-berlin.de/~wwwgravi

REFERENCES

1. Götze, H.-J. and the MIGRA - group, 1996: Group updates the gravity data base in the Central Andes (20° - 29° S). EOS Transactions, American Geophysical Union, in print.
2. Götze, H.-J., M. Schmitz, P. Giese, S. Schmidt, P. Wigger, G. Schwarz, M. Araneda, G. Chong D. and J. Viramonte 1995: Las estructuras litosféricas de los Andes Cantrales australes basados en interpretaciones geofísicas. IGCP Special Volume, Revista Geológica de Chile, Vol 22, No. 2, 179-192.
3. Coira, B., Davidson, J., Mpodozis, C. and Ramos, V.A., 1982. Tectonic and magmatic evolution of the Andes of northern Argentina and Chile. Earth Sci. Rev., 18, pp. 303-332.
4. Reutter, K.-J., E. Scheuber and P. Wigger (Editors) 1994. Tectonics of the Southern Central Andes, Springer Verlag, Heidelberg, pp. 333.
5. Isacks, B.L., Uplift of the Central Andean Plateau and Bending of the Bolivian Orocline. Journ. Geophys. Res., Vol. 93, No. B4, 3211 - 3231, 1988.
6. Wigger, P.J., M. Schmitz, M. Araneda, M., G. Asch, S. Baldzuhn, P. Giese, W.-D. Heinsohn, E. Martinez, E. Ricaldi, P. Röwer and J. Viramonte 1994: Variation of the crustal structure of the Southern Central Andes deduced from seismic refraction investigations. In: Tectonics of the Southern Central Andes (Eds.: Reutter, Scheuber, Wigger), Springer Verlag Heidelberg, pp. 23-48 .
7. Schmitz, M., 1994. A balanced model of the Southern Central Andes. TECTONICS, 13, no.2, pp. 484-492.
8. Cahill, Tom (1990): Earthquakes and Tectonics of the Central Andean Subduction Zone.- PhD Thesis, Cornell University, Ithaca, New York.

9. Whitman, D., Isacks, B.L., Chatelain, J.-L., Chiu, J.-M., and Perez, A.,1992. Attenuation of high-frequency seismic waves beneath the Central Andean Plateau. J. Geopgys. Res., 97 (B13), pp 19929 - 19947.

10. Götze, H.-J., Lahmeyer, B., Schmidt, S. and Strunk, S. 1994: The Lithospheric Structure of the Central Andes (20-26°S) as inferred from Quantitative Interpretation of Regional Gravity. In: Tectonics of the Southern Central Andes (Eds.: Reutter, Scheuber, Wigger), pp. 7-21, Springer Verlag Heidelberg.

Proc. 30ᵗʰ Int'l. Geol. Congr., Vol. 4, pp. 143-152
Hong Dawei (Ed)

Receiver Functions and Tibetan Lithospheric Structure

WU QINGJU   ZENG RONGSHENG

Institute of Geophysics, State Seismological Bureau, Beijing 100081, *China*

Abstract

Broadband receiver functions derived from teleseismic P waveforms recorded on 11 PASSCAL instruments during the Sino-US joint project " Tibetan Lithospheric Structure and Geodynamics" are inverted for the vertical S wave velocity structure beneath the recording sites in the Tibetan plateau. The receiver functions are isolated by high-resolution maximum entropy deconvolution in time-domain, which are more precise than those in frequency-domain. The mean receiver function is obtained by stacking a set of teleseismic events clustering in a small range of backazimuth and epicentral distance to increase the signal-to-noise ratio. Kennett complete seismogram and Randall's efficient differential seismogram are utilized in waveform modeling. Time-domain waveform inversion is implemented by "jumping" algorithm. The model smoothness constraint is included in waveform inversion procedure. The inversion result reveals that there exists an obvious Moho offset near Bangong suture, which may indicate the collisional front of Indian crust injecting into Eurasian lower crust.

Key words: Receiver Function, Tibetan Plateau, Waveform Inversion, S Wave Velocity Structure.

INTRODUCTION

The Tibetan plateau is one of the most prominent and peculiar tectonic unit over the world, and has long been the focus of geoscience. However, the details of the collisional process and the uplift mechanism remain controversial. The most popular uplift model includes (1) the underthrust model and its variation [1-6], and (2) the crustal shortening and thickening model [7]. It is still difficult to discriminate different uplift models due to insufficient information on the lithospheric structure of the Tibetan plateau.

S wave velocity is an important geophysical parameter of the lithospheric structure, and is difficult to be obtained by other methods. Teleseismic P waveform provides an excellent approach to get S wave velocity, due to the fact that it consists of P-to-S converted waves and reverberations generated by crustal discontinuities beneath the seismic station. The teleseismic P wave arriving at the receiver structure can be simplified as a plane wave with steep incident angle, and samples the local structure with relatively small horizontal extent, accordingly, the teleseismic P waveform can be approximately modeled by the plane wave propagation in the horizontal stratified media beneath the recording site.

The teleseismic P waveform itself contains information related to source time history, focal mechanism, near-source structure, mantle propagation effects, near-receiver structure, and instrument response, the source and mantle propagation effects should be eliminated from it in order to delineate the receiver structure with the waveform data. Source equalization scheme is proposed to recover receiver function from the complicated teleseismic P waveform. Receiver function is isolated from the teleseismic P waveform through the deconvolution of vertical component from the radial and tangential components after the vector rotation of horizontal ones [8,9]. It is generally recognized that the receiver function is just related to local receiver structure, with the source and mantle propagation effects removed. It is particularly sensitive to the sharp variation of S wave velocity.

METHOD

Three-component teleseismic P waveform can be considered as the convolution of three independent factors [8], namely,

$$D_Y(t) = I(t) \cdot S(t) \cdot E_Y(t)$$
$$D_R(t) = I(t) \cdot S(t) \cdot E_R(t)$$
$$D_T(t) = I(t) \cdot S(t) \cdot E_T(t) \tag{1}$$

where $S(t)$ represents the equalized source time function consisting of the overall effects from the source to the investigating depth beneath the recording station, $E(t)$ the receiver structure response, in which subscript R, T, R corresponds to the radial, transverse and vertical direction respectively, and $I(t)$ instrument response. Real data observation and theoretical calculation have confirmed that the vertical response of teleseismic P waveform consists of the major direct arrival and minor second arrivals, thus it can be approximated by Dirac function, i. e.

$$E_Y(t) \approx \delta(t) \tag{2}$$

Under such assumption, the vertical component is nothing but the factor which should be removed from the teleseismic P waveform,

$$I(t) \cdot S(t) \approx D_Y(t) \tag{3}$$

The source equalization scheme is conducted through the deconvolution of vertical component from the horizontal components. The resulting time series is named the receiver function, which is dominated by the horizontal response of receiver structure for the teleseismic P wave, and is almost irrelevant to source and mantle propagation effects. The conventional deconvolution is carried out in frequency domain through the spectrum division, expressed as the following:

$$E_R(\omega) = \frac{D_R(\omega)}{I(\omega)S(\omega)} \approx \frac{D_R(\omega)}{D_V(\omega)}$$

$$E_T(\omega) = \frac{D_T(\omega)}{I(\omega)S(\omega)} \approx \frac{D_T(\omega)}{D_V(\omega)} \tag{4}$$

Owing to the band-limited data and random noise, the near-zero spectrum of vertical component will cause the spectrum division unstable, the water-level technique is introduced to keep deconvolution stable [10]. On the other hand, it will also result in low resolution of receiver function. In order to overcome the inherit defect in frequency-domain deconvolution, two time-domain deconvolution, i. e. Wiener deconvolution and maximum entropy deconvolution have been developed, synthetical and observational data show that both time-domain deconvolution methods are excellent[11].

After the receiver functions are recovered from the selected teleseismic events clustering within a small range of backazimuth and epicentral distance, a set of receiver functions are stacked to increase the signal-to-noise ratio and enhance the coherent P-to-S converted phases and reverberations. The resulting receiver function represents the average horizontal response of the receiver structure. Moreover, the stacking procedure also provides statistical upper and lower bounds of the mean receiver function to evaluate the data quality and waveform modeling.

It has been demonstrated that the receiver function is a scaled version of the radial component of displacement with the removal of the P multiples [12]. The receiver function, although developed from P waveform, is most sensitive to the S wave velocity structure beneath the station, since P-to-S converted waves dominate the horizontal components[9].

The teleseismic P waveform inversion is to decode a simplified velocity model of the receiver structure from the receiver function. The synthetical receiver function may be described by

$$d_j = R_j[m] \qquad j = 1,2,3,\cdots,N \tag{5}$$

where d_j represents the jth data, and R_j represents the functional which operates on the model m to produce the receiver function.

The relationship (5) is nonlinear and computed by Kennett complete seismogram in our study [13].

Because receiver function is most sensitive to the vertical variation of S wave velocity, the velocity structure is parameterized as M-dimensional vector of S wave velocities with fixed layer thickness, P wave velocity is dynamically adjusted by assuming the empirical relationship $\alpha = 1.73\,\beta$, and densities is also approximated throughout the model using petrologic statistical expression $\rho = 0.32\,\alpha + 0.77$, where ρ, α and β represent density, P and S wave velocity respectively.

Generalized linearized inversion solves for an estimate of the true velocity model m in an iterative manner, with an initial approximation to the local receiver structure which is close to the true velocity model m. Under such conditions the problem may be linearized by expanding the observed receiver function in a Taylor series about starting model m_0.

$$R_j[m] = R_j[m_0] + \langle D,\ \delta m \rangle_j + O \parallel (\delta m^2) \parallel \qquad (6)$$

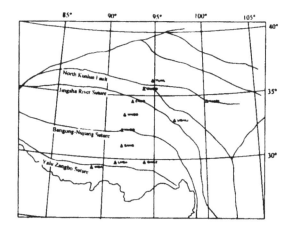

Figure 1. The location of PASSCAL seismometers (triangle) in the Tibetan plateau.

The matrix D is $N \times M$ dimension, and is often referred to as the data kernel. The i, k th element of D is $\partial\,R_i[m]\,/\partial m_k$, representing the sensitivity of ith time sample in the synthetic receiver function to small changes in the kth layer velocity. The columns of matrix D constitute differential seismograms demonstrating the relative variation of waveform with layer velocity, which is computed by the efficient complete differential seismogram technique of Randall [14].

Equation (6) is used to directly solve for model vector m itself rather than the model correction vector δm, and is termed "Jumping" algorithm, which can decrease nonuniqueness to certain extent and obtain reasonable model estimate. When "jumping" algorithm is implemented in waveform inversion procedure, the smoothness constraint can be exerted on waveform inversion through minimizing the model roughness norm [15], to dampen the large amount of rapid variation of velocity with depth[16].

DATA AND RESULTS

From July, 1991 to July 1992, Sino-US joint project on "Tibetan Lithospheric Structure and Geodynamics" was conducted [2, 17], and 11 broadband digital PASSCAL seismometers were deployed within the Tibetan plateau. Fig. 1 presents the location of PASSCAL seismometers. More than 300 teleseismic events ranging from 35° to 90° in epicentral distance were recorded during the experiment. It provides an excellent opportunity for utilizing teleseismic P waveform to study the Tibetan lithospheric structure, particularly the local S wave velocity beneath the recording stations. A preliminary study of the receiver functions at three stations (TUNL, WNDO, and XIGA) had revealed the great thickness of the crust and the difference in velocity structures between the north-central and southern Tibetan plateau [18]. The inversion results of receiver functions for total 11 broadband stations are presented here.

About 70 teleseismic events with high signal-to-noise ratio, from northeastern backazimuth, with an average epicentral distance about 40°, are selected for our investigation of the Tibetan crustal velocity structure. The P waveforms for the selected events are cut to a length of 120s within the window of 20s pre-event and 100s post-event , which contain the reverberation phases from the Moho. The mean values are also removed from the teleseismic P waveform data set, and the raw data are rotated into theoretical radial and tangential components.

The maximum entropy deconvolution are utilized to isolate the receiver functions from the total 70 teleseismic events recorded on the 11 PASSCAL stations respectively, then those receiver functions with strong coherence are selected for stacking to obtain the average receiver functions for each station respectively. The stacking gathers include at least 10-20 individual teleseismic events for each station.

The receiver structures are parameterized into one-dimensional S wave velocity model with a thickness of 2 km, due to the frequencies higher than 1 Hz being filtered in receiver functions by a Gaussian filter with coefficient of 2. 5. The generalized linearized inversion approach is applied for the receiver functions. The initial model is determined on the basis of deep seismic sounding and local seismic phases as well as body waves. The smoothness norm is included to make a trade-off between model roughness and waveform fitness. Kennett seismogram and Randall's differential seismogram are applied in computing synthetical seismogram to obtain the complete response in horizontal receiver structure, and S wave velocity distribution is obtained for each station.

Fig. 2 presents the receiver function and its waveform inversion result for WNDO station. The result should be reliable, judging from the small deviation among the

148

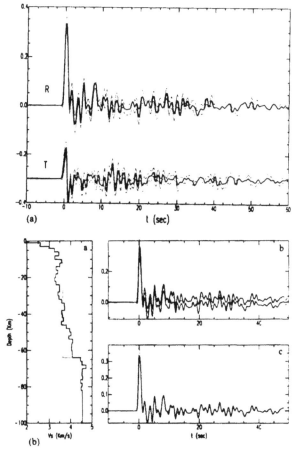

(a)   t (sec)

(b)   Vs (Km/s)   t (sec)

Figure 2. (a) The average receiver functions (solid
line) and lower and upper bounds (dash line) for WN-
DO station (R - radial, T - tangential). (b) The waveform inversion result for receiver function in WNDO, a
- S wave velocity model (dash line - initial model, solid
line - inversion model), b - the evaluation of waveform
modeling (dash line - synthetical waveform, solid line -
the lower and upper bounds of observational data), c -
the fitness of waveform modeling (dash line - synthetic
waveform, solid line - observational data).

individual receiver func-
tions. The crustal struc-
ture here seems lateral homogeneous due to the low-
value of transverse component. The P-to-S convert-
ed wave from Moho is
clearly detected about 8. 2s
after the direct P wave.
Waveform inversion of re-
ceiver function indicates
that Moho is 64km deep
beneath WNDO. S wave
velocity increases gradually from 3. 7km/s at 50km
deep to 4. 4km/s at Mo-
ho. Another significant
phenomenon is noted that
a low velocity zone of
10km thick occurs in the
upper crust.

The receiver functions and
their waveform inversion
results for each station are
not discussed respectively,
but summarized together
as the follows.

The waveform fitting sta-
tus of the receiver func-
tions for the 11 recording
stations is shown in Fig.
3. According to Fig. 3,
the P-to-S converted
waves from Moho can be clearly detected, and the receiver functions are excellent-
ly modeled. It should be pointed out that significant difference also exists among
those receiver functions recorded on different stations, which indicates that the Tibetan lithospheric structure is lateral inhomogeneious within the studying range.

The S-wave velocity models for the 11 seismic stations deduced from the teleseismic P waveform are aligned in two profiles (a and b) from south to north (Fig. 4). It reveals the major features of the Tibetan lithospheric structure.

In the inversion model, the Moho, the major interface, is determined at the depth where S wave velocity reaches 4. 4-4. 6km/s, and is characterized by first-order discontinuity or velocity gradient zone. S wave velocity of Moho inferred from receiver function is consistent to the travel-time analysis of local Sn wave [5].

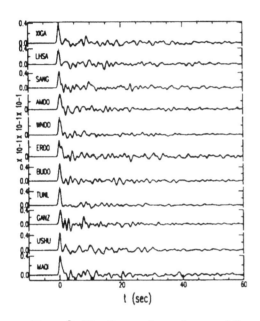

t (sec)

Figure 3. The fitness of waveform modeling for the Tibetan receiver functions (solid line - observational data, dash line - synthetical waveform), stations are also labeled in the front of each channel.

The depth of Moho is about 60-80km, and exists obvious relief in the Tibetan plateau. Moho dips northward slightly south to Bangong suture, reaching 84km deep at AMDO, maybe the deepest site. It becomes significantly shallow north to Bangong suture, about 64km deep at WNDO, then increases gradually to the north, and reaches 78km deep at BUDO. Moho is about 64km deep at TUNL.

The low velocity zone seems not to be a general feature in the Tibetan crust, but exists in the upper or lower crust beneath certain stations. A low velocity zone is located in the upper crust at SAGN, AMDO, ERDO, and BUDO, and its lower interface forms a significant discontinuity dipping northward in the Tibetan upper crust. Another low velocity zone about 10km thick occurs at the depth of 50-60km in the lower crust at XIGA and SANG. It is noted that a low velocity zone at least 20km thick also occurs at the depth of 40-60km at BUDO.

Several intracrustal discontinuities can be identified in the Tibetan crust. The discontinuity h_1 varies from 18km deep at XIGA to 30km deep at BUDO with an evident velocity jump, noticeably dipping northward and representing the bottom of low velocity zone at SANG, AMDO, ERDO, and BUDO. The discontinuity h_2 appears at XIGA about 50km deep and gradually decrease to 40km deep at BUDO, forming the top of low velocity zone at XIGA, SANG, and BUDO, and also constituting the boundary of velocity gradient zone at AMDO and ERDO. The discontinuity h_3 is located at 60km deep from XIGA to SANG, forming the bottom of low velocity zone at XIGA and SANG, but disappears north to Bangong suture. The intracrustal discontinuities in the Tibetan plateau had been confirmed by wide-angle reflections [19-25].

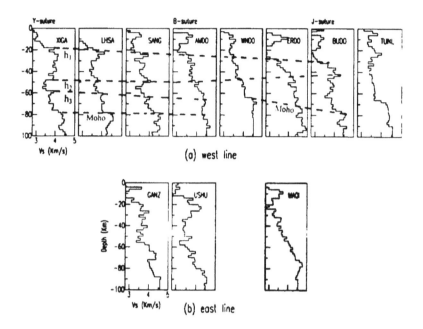

Figure 4. The crustal velocity model inferred from receiver functions in the Tibetan plateau, stations are represented in each velocity model, major interfaces are also indicated, Y suture - Yalu Zangbo Suture, B suture - Bangong Suture, J suture - Jingsha River suture

A significant result is that there exists Moho offset with the depth variation of 15km across Bangong suture, Moho becomes shallow across Bangong suture suddenly, and S wave velocity increases monotonously from 3.5km/s at 23km deep to 5.0km/s at 85km deep. The phenomenon that Moho becomes shallow north to Bangong suture has also been detected by DSS investigation [24, 25], surface wave dispersion [26], and local phases [5,6], especially the latest study on local Pn travel-time data with the velocity model from our receiver function results as the constraint [27]. The consistence of various data and methods confirms the Moho offset across Bangong suture reliable.

CONCLUSIONS

In our study, about 70 teleseismic events clustering in a small range of backazimuth and epicentral distance with high S/N ratio are selected from the northeastern direction, high-resolution maximum entropy deconvolution and staking procedure with at least 10-20 individual events keep the Tibetan receiver functions much reliable, the consistence of waveform inversion results from receiver functions with different data and methods makes our result confident.

In the Tibetan plateau, Moho is about 60-80km deep, with obvious undulation; several obvious S wave velocity discontinuities exist in the Tibetan crust; the low velocity zone only exists in the upper or lower crust at certain sites; a Moho offset with the depth variation 15km is close to Bangong suture, Moho is deeper to the south of it, and shallower to the north; based on the Moho offset near Bangong suture and the velocity gradient zone under AMDO station, the material mixing process between Indian and Eurasian crusts is suggested during the Indian crust injecting into the lower crust of Eurasian continent.

Acknowledgements

The first author is indebted to Prof. Feng Rui for his helpful comments about this study and careful review on the manuscript, thanks are also given to Ding Zhifeng and Wu Jianping for their help in this study. This work has been supported by National Foundation of Sciences.

REFERENCES

1. W. L. Zhao and W. J. Morgan. Injection of Indian crust into Tibetan lower crust; a two-dimensional finite element model study, Tectonics 6, 489-504 (1987).
2. R. S. Zeng, F. T. Wu, and T. J. Owens. An introduction to the Sino-US joint project "Lithospheric Structure and Dynamics in Tibetan Plateau", Acta Seism. Sinica 6(2), 249-250 (1993).
3. R. S. Zeng, J. S. Zhu, B. Zhou, Z. F. Ding, Z. Q. He, L. P. Zhu, L. Xun, and W. G. Sun. Three-dimensional seismic velocity structure of the Tibetan plateau and its eastern neighboring area with implications to the model of collision between continents, Acta Seism. Sinca 6(2), 251-260 (1993).
4. R. S. Zeng and W. G. Sun. Seismicity and focal mechanism in Tibetan plateau and its implications to lithospheric flow, Acta Seism. Sinica 6(2), 261-287 (1993).
5. R. S. Zeng and Y. H. Ming. Regional seismic phases and lithospheric structure in Tibetan plateau, Submitted to Acta Seism. Sinica (1996).
6 R. S. Zeng, Z. F. Ding, and Q. J. Wu. A review on the lithospheric structures in the Tibetan plateau and constraints for dynamics, PAGEOPH 145, 425-443 (1995).
7. J. F. Dewey and J. M. Bird. Mountain belts and the new global tectonics, J. Geophys. Res. 75, 2625-2647 (1970).
8. C. A. Langston. Structure under Mount Rainier, Washington, inferred from teleseismic body waves, J. Geophys. Res. 84, 4749-4762 (1979).
9. T. J Owens, G. Zandt, and S. R. Taylor. Seismic evidence for an ancient rift beneath the Cumberland Plateau, Tennessee; A detailed analysis of broadband teleseismic P waveforms, J. Geophys. Res. 89, 7783-7795 (1984).
10. R. A. Wiggins and D. V. Helmberger. Upper mantle structure of the western United States, J. Geophys. Res. 78, 1870-1880 (1973).
11. Q. J. Wu. *Waveform Inversion from Teleseismic Events and Lithospheric Structures in Tibetan Plateau*, Ph. D. thesis, Institute of Geophysics, State Seismological Bureau of China, Beijing (1996).
12. C. J. Ammon. The isolation of receiver effects from teleseismic P waveforms, Bull. Seismol. Soc. Am. 81, 2504-2510 (1991).
13. B. L. N. Kennett. *Seismic Wave Propagation in Stratified Media*, Cambridge University Press, New

York (1983).

14. G. E. Randall. Efficient calculation of differential seismograms for lithospheric receiver functions, Geophys. J. Int. 99, 469-481 (1989).

15. S. C. Constable, R. L. Parker, and C. G. Constable. Occam's inversion: a practical algorithm for generating smooth models from electromagnetic sounding data, Geophysics 52, 289-300 (1987).

16. C. J. Ammon, G. E. Randall and G. Zandt. On the nonuniqueness of receiver function inversions. J. Geophys. Res. 95, 15,303-15,318 (1990).

17. T. J. Ownes, G. E. Randall, F. T. Wu, and R. S. Zeng. PASSCAL Instrument performance during the Tibetan plateau passive seismic experiment, Bull Seismol. Soc. Am. 83(6), 1959-1970 (1993).

18. L. P. Zhu, R. S. Zeng, F. T. Wu, T. J. Owens, and G. E. Randall. Preliminary study of crust-upper mantle structure of the Tibetan plateau by using broadband teleseismic body waveforms, Acta Seism. Sinica 6(2), 305-316 (1993).

19. J. W. Teng, K. Z. Sun, S. B. Xiong, Z. X. Yin, H. Yao, and L. F. Chen. Deep seismic reflection waves and structure of the crust from Dangxung to Yadong on the Xizang plateau (Tibet), Phys. Earth Plant. Int. 31, 293-306 (1983).

20. J. W. Teng, S. B. Xiong, Z. X. Yin, Z. X. Xu, X. J. Wang, D. Y. Lu, G. Jobert, and A. Hirn. Structure of the crust and upper mantle pattern and velocity distributional characteristics at northern region of the Himalayan Mountains, Acta Geophys. Sinica. 26, 525-540 (1983).

21. J. W. Teng, S. B. Xiong, X. Z. Yin., X. J. Wang, and D. Y. Lu. Structure of the crust and upper mantle pattern and velocity distributional characteristics in the northern Himalayan Mountain region, J. Phys. Earth 33, 157-171 (1985).

22. J. W. Teng, X. Z. Yin, and S. B. Xiong. Crustal structure and velocity distribution beneath the Serlin Co-Peng Co-Naqu-Suo county region in the northern Xizang (Tibet) plateau, Acta Geophys. Sinica. 28 (Suppl.), 28-42 (1985).

23. J. W. Teng. Explosion study of the structure and seismic velocity distribution of the crust and upper mantle under the Xizang (Tibet) plateau, Geophys. J. R. Astron. Soc. 89, 405-414 (1987).

24. S. B. Xiong, J. W. Teng, and Z. X. Yin. The thickness of the crust and undulation of discontinuity in Xizang (Tibet) plateau, Acta Geophys. Sinica, 28 (Suppl), 16-27 (1985).

25. D. Y. Lu and X. J. Wang. The crustal structure and deep internal processes in the Tuotuohe-Golmud area of the north Qinghai- Xizang plateau, Bull. Chinese Academy of Geol. Sci. 21, 227-337 (1990).

26. C. Brandon and B. Romanowicz. A "no-lid" zone in the central Chang Thang platform of Tibet; evidence from pure path phase velocity measurements of long period Rayleigh waves, J. Geophys. Res. 91, 6547-6564 (1986).

27. R. S. Zeng, Z. F. Ding, and Q. J. Wu. Lateral variation of crustal structure from Himalaya to Qilian and its implication on continental collision process, in "Structure of the Lithosphere and Deep Process", Proc. 30th IGC (1996).

Proc 30ᵗʰ Int'l. Geol. Congr., Vol. 4, pp. 153-160
Hong Dawei (Ed)
© VSP 1997

Lithospheric Isostasy, Flexure and the Strength of Upper Mantle in Tibetan Plateau

Yong WANG[+] Rui FENG[++] Houtse HSU[+]

Institute of Geodesy and Geophysics, CAS, Wuhan, 430077, P. R. China, E-mail:ywang@asch.whigg.ac.cn

Institute of Geophysics, SSB, Beijing, 100081, P. R. China

Abstract

Based on Airy-Heiskanen model of local compensation, a new version of the isostatic residual gravity map with 10' × 10' resolution over Tibetan Plateau was presented using new gravity data, which offer significant additional information in understanding the nature of tectonic processes, tectonic zonation, and the investigation of the stress. We estimated the elastic strength of the lithosphere beneath the Tibetan Plateau and its vicinity using coherence techniques, and obtained an effective elastic thickness of about 85 km. We calculated the isostatic residual geoid over Tibetan Plateau. Comparisons of the isostatic residual geoid at intermediate wavelength with the geoid predicted by density anomalies in upper mantle inverted from seismic velocity anomalies in a viscous Earth with different viscous structure of upper mantle, the low viscosity in the uppermost mantle seems to be required beneath Tibetan Plateau, which indicated the existence for a hot uppermost mantle beneath Tibet.

Keywords: Isostatic Residual Gravity, Tibetan Plateau, Flexure, Strength, Lithosphere, Upper Mantle

INTRODUCTION

The Tibetan Plateau is one of the most prominent and peculiar geological structure in the continent. It is the highest and largest plateau of the world with an average height of 4500m. The Tibetan Plateau, the Himalaya and the Karakoram are the most spectacular consequences of the collision of the Indian subcontinent with the rest of Eurasia in Cenozoic time. The deep structure of the Himalaya and Tibet is poorly known, compared with that of much of the world, but by the same comparison, its geological study has a tradition of importance exceeded by few areas. The concept of isostasy, which perhaps represents the birth of geophysics as subdiscipline of and a contributor to the geological sciences, grew from geodetic work in India and from the realization that the deep structure of the Himalaya and Tibet had contributed to large discrepancies in surveyed position.

The variations of the Earth's gravitational attraction reflect how much the Earth's density structure departs from hydrostatic equilibrium and what mechanisms maintain such departures. The principle of isostatic equilibrium is an important theory to investigate the density anomalies and mechanical behavior of the lithosphere. These allow us to accept a hypothesis that the sum of anomalous masses along any segment of the Earth's radius between its surface and the subsurface at which hydrostatic conditions are reached equals to zero. These conditions are those of classical isostasy, Pratt's and Airy's model[11, 1] are particular cases of classic isostasy. Wang [17] published the first isostatic gravity anomalies map of China with the resolution of $1° \times 1°$. To examine whether the Himalaya are in local isostatic equilibrium, Lyone-Caen and Molar [7, 8] compared measured Bouguer gravity anomalies with those calculated from the corresponding average topographic profiles, assuming local isostatic equilibrium. The isostatic gravity map published by Wang [17] based on poor resolution of the gravity and topography data. The Bouguer gravity anomalies usually obtained from the simple Bouguer correction. In Tibetan Plateau and Himalaya, the terrain correction can be enormous up to 100 mgal, and can not be ignored. At present, a new version of the isostatic gravity map of Tibetan Plateau has become possible, because of the addition and compilation of a gravity data base. The new isostatic gravity map can display more clearly the gravity anomalies caused by geologic bodies in the upper parts of crust, and offer the significant additional information in understanding the nature of tectonic processes, tectonic zonation, and the investigation of stress state of the Earth's interior.

One weakness to Pratt and Airy local compensation model is that the crustal and lithospheric strength have been ignored. At long wavelength, the gravity field arises from the effects of mass anomalies in the mantle and the warping of density interfaces(such as core-mantle boundary) in response to the flow driven by the buoyancy of these anomalous masses. At shorter wavelength, the gravity field is sensitive to the strength of the Earth's lithosphere supporting loads near or on the surface, such as topography. In this study, we investigate the flexure of lithosphere using short wavelength gravity anomalies, and the strength of upper mantle using intermediate wavelength ones.

ISOSTATIC RESIDUAL GRAVITY MAP OF TIBETAN PLATEAU

Gravity values for the new map come from the new gridded data set. This grid has a $10' \times 10'$ interval and contains the complete Bouguer gravity values (Fig. 1). For Tibetan Plateau areas, approximately 60% of all $10' \times 10'$ cells have at least one gravity observation available. We use the least square collocation (LSC), which is the only known geodetic technique in using heterogeneous data, to interpolate the gravity values for the lack of gravity observation areas. In the practical use of LSC, we exploded the so-called remove/restore procedure with the Institute of Geodesy and Geophysics, CAS, 1993 local gravity model(IGG93C) [4] to degree 360 as the reference field. The error parts of the covariance functions are computed using the error estimates of IGG93C coefficients in a legendre polynomial series. A LSC computation is carried out in $1° \times 1°$ cell. For the LSC computation, see the work of Tscherning [16]. We chose to use an Airy-Heiskanea model with local compensation. This model requires a choice of values for three parameters: a depth of compensation for sea level elevation d_s, a density contrast across the bottom of the root $\Delta\rho$ and a density of the topographic load ρ_t. These parameters determine the depth to the bottom of the root under land areas by the relation

$$d = d_s + e(\rho_t / \Delta\rho) \qquad (1)$$

Where e is the elevation of the topographic surface. For the Airy-Heiskanen parameters we chose a depth to the bottom of the root of 30.0 km for sea level elevations, a density contrast across the root of $0.35g/cm^3$, and a topographic density of $2.67g/cm^3$.

The attraction of the Airy-Heiskanen root at a point on the Earth's surface was calculated in two steps: the attraction on a flat earth out to a distance of 166.7 km was calculated by using a program

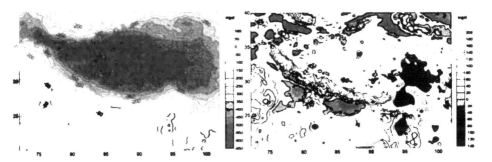

Figure 1. Bouguer gravity anomaly map   **Figure 2.**. New version isostatic residual gravity map

[13] based on the fast Fourier transform algorithm developed by Parker [10]. This result was combined with a published regional field for the attraction of both topography and root beyond 166.7 km to a distance of 180 on a spherical earth [5]. A mismatch exists between the model parameters for the published results of Karki(sea level compensation depth, 0 km; density contrast, $0.6g/cm^3$), and those used for calculation of the root attraction inside 166.7 km (30 km and $0.35g/cm^3$). The problem has been discussed by Simpson [14]. For the Tibetan Plateau areas, where a broad area has an average elevation of almost 4 km. A mismatch of as much as 10 mgal may occur. The final step in the production of the isostatic residual gravity map was subtraction of the isostatic region field from the Bouguer gravity map. The isostatic residual gravity values are subject to error caused by uncertainly in station elevation, terrain correction, and gravity measurement. We expect the most of the isostatic residual gravity values are accurate to better than 10 mgal.

Figure 2 shows the new version isostatic residual gravity map over Tibetan Plateau. It can be used to investigate the correlation between geology and isostatic residual gravity, and offer significant additional information in understanding the nature of tectonic processes, tectonic zonation, which will be discussed in another paper. In here, we simply describe only the fundamental features.

The new isostatic residual gravity map shows the change range of isostatic residual gravity in Tibetan Plateau and its vicinity reaches 300 mgal. The maximum positive anomalies along Pamir-Karakoram-Himalaya are as much as 150 mgal. The maximum negative anomalies over Fergana basin, Tarim basin, front depression of Kunlun and Parim mountain are as much as -150 mgal. There are a negative anomaly in the Indian plain. The isostatic equilibrium prevails in Tibetan Plateau interior, where isostatic residual gravity are less then 40 mgal. There are a little different between the old isostatic residual gravity map with $1°×1°$ resolution and the new one. The magnitude of gravity anomalies on new map is larger than that of gravity anomalies on old map. This is because the new map has higher resolution than the old one that smooth the isostatic residual gravity anomalies. On the other hand, the new maps display more clearly the gravity anomalies caused by geologic bodies in the upper parts of the crust.

LITHOSPHERIC FLEXURE

There are two primary types of models of the isostatic compensation of topographic features. In the first, compensations occur directly beneath the topography by thickening a constant density crust or by lateral changes in density of crust. This type of model of the isostatic compensation have been mentioned in last part, and is used to prepare a new isostatic residual gravity map. In the second type of model, the flexure model, loads are partially supported by elastic stresses within the lithosphere plate overlying a weak fluid asthenosphere, and compensation occurs on a regional basis. In the flexure model, the response of the plate is often characterized by the flexure rigidity(D) or equivalently, by the effective elastic thickness(T_e). Actually, the Airy model does not consider the lithospheric strength, and correspond to a flexure rigidity of zero.

Method for estimating the rigidity of lithosphere from gravity data with the context of a purely elastic plate model fall into two categories: direct forward modeling of the data, in which one compares the observed Bouguer gravity to that predicated by various models of plate flexure, and spectral admittance techniques, in which one compares only the average rations of Bouguer gravity to topography in selected wavelength ranges to those predicted by plate models, which the assumption inherent in the admittance approach require that all loads deforming the elastic plate be topographically expressed and that the Earth's response to loading be stationary in space. While these conditions are usually met for midplate loading in the ocean basins, they rarely hold on the continents where significant subsurface mass anomalies may load the plate. For this reason, Foryth

[6] introduced the coherence function as a robust means of estimating Te in continental region. The coherence method assumes that topographic relief reflects both surface and surface density contracts, or load , and exploits the relationship between the Bouguer, gravity and topography as a function of the wavelength. Elastic plate thickness for Himalaya, Kunlun are also derived from forward modeling of gravity data as described in the references cited. In this study , we use the coherence [3] method to examine Te using new gravity date. The coherence function is defined as

$$\gamma^2 = \frac{<H_T W_T + H_B W_B>^2}{<H_T^2 + H_B^2><W_T^2 + W_B^2>} \tag{2}$$

Where H is the surface topography , W is the relief at the Moho , and subscripts t and b stand for top (or surface topography) and bottom (or base of the crust) loads , respectively . Equation (2) is simplified by assuming that surface and subsurface loads are independent processes . The Te values obtained from the best fit mode of computed coherence and from forward modeling of gravity data for Tibetan Plateau and its vicinities are listed in Table 1.

Table 1. Effective Elastic Thickness of Lithosphere Beneath Tibetan Plateau and Vicinity

Region	T_e (km)	Method	f	Reference
Tibet (Himalaya)	80-90	Coherence	1.0	this study
Himalaya (India Shield)	90±10	Forward		Lyon-Caen and Molnar [7]
Pamir	20-25	Coherence	0.4	this study
Pamir (Tadjik Depression)	15±5	Coherence	0.2	McNutt et. al. [9]
Tarim	40-50	Coherence	1.5	this study
Kunlun (Tarim Basin)	40	Forward		Lyon-Caen and Molnar [7]
Tien Shan (Tarim Basin)	40±20	Coherence	1.5	McNutt et. al. [9]

On the basis of coherence techniques, we would predict an elastic plate thickness beneath the Tibetan Plateau and its vicinity. The estimate obtained here are consistent with the values from forward modeling in Table 1. Our estimates, however, are a little greater than that of Lyon-Caen and Molnar. In Table 1, there is an important and significant factor f, which is the ratio of internal/surface loading. For the Himalaya, the ration f is equal to 1, which indicate there are almost the same amount of surface and subsurface load. The Tien Shan mountain have a much small amount of subsurface load since f is less than 1.0. For Tarim, the effective elastic thickness Te is greater than 1.0 , that show the lithosphere underthrusting the Pamir is very weak, and is dominantly loaded by forces other than the topography.

STRENGTH OF UPPER MANTLE

In order to investigate the rheological behavior of continental upper mantle using Earth's gravity, It would be helpful to remove the gravity disturbance caused by surface topography [2]. The Earth's gravity field are of two types: a Earth's gravity anomalies, and another geoidal anomalies. Kogan and McNutt [6] discussed the Earth's gravity field over northern Eurasia and variation in the strength of the upper mantle. In their study, gravity anomalies were used, and the gravity effects due to topography and its compensation had not been subtracted from the observed gravity anomalies. Actually, The geoidal anomalies are more suitable for the deep tectonic processes. One reason is its stability. In addition, geoidal anomalies resemble the potential in having $1/r$ dependence with distance from their source, in contrast to the $1/r^2$ dependence of gravity. In this study, therefore, we first remove from the geoid the effects of topography compensated, and use the intermediate wavelength geoidal anomalies to study the strength of upper mantle beneath Tibetan Plateau. Since we attempt only a regional analysis for mantle rheology behavior beneath Tibetan Plateau and its vicinity. We removed spherical harmonic degree 1 through 6 from geoidal data according to the Kogan and McNutt's idea. Figure 3a shoes the isostatic residual geoid at intermediate wavelength.

The relation between a thin sheet of anomalous masses $\delta\sigma_{lmi}$ of degree l, order m located on the i th spherical shell of the Earth and the gravity anomaly, it produces at the Earth's surface, δg_{lmi} is [12]

$$\delta g_{lmi} = \frac{4\pi\gamma(l-1)}{2l+1} G_{li} \delta\sigma_{lmi} \tag{3}$$

where r is Newton's gravitational constant , and G_{li} is an array of dimensionless Love number. The Love number include the direct gravitational effect of the mass sheet as well as contributions to the gravity field from the warping of density interfaces caused by the convection driven by the anomalous mass. To obtain G_{li} , one must solve the questions of continuity and motion given a set of boundary conditions, a viscosity model, and a constitutive relation between stress and train. In this study, we adopt the s-wave anomaly model of Su et al. [15] in term of spherical harmonic expression to degree and order 12 , and remove also first six spherical harmonic coefficients, and convert to be space domain., and use Birch's law to convert the velocity anomalies into density anomalies.

We predicted the geoidal anomalies three viscosity model. M_1 Model had a uniform viscosity of upper mantle and lower mantle. M_2 Model had a uniform viscosity upper mantle ten times less viscous than the lower mantle. In M_3 Model , the viscosity was decreased one more order of magnitude between the depths of 100 and 400 km. Figure 3 a display the predicted geoidal anomalies for three different structure of the upper mantle. comparison of the observation geoidal anomalies with the predicted one, show that the correlation between the observation geoid anomalies and the at predicted by M_1 and M_2 model is negative, and the coherence is low. For M_3 model, the opposite is true, the correlation is positive and coherence is high, that show Model $_3$ provides the best fit for the Tibetan Plateau.

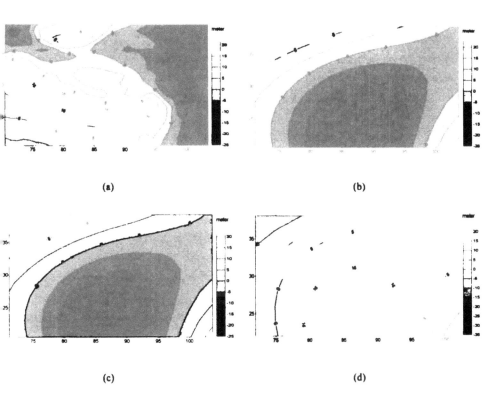

(a) (b)

(c) (d)

Figure 3. Isostatic residual geoid at intermediate wavelength over Tibetan Plateau and its vicinity. (a) observed geoid, (b) predicted geoid by M_1 model, (c) predicted geoid by M_2 model, (d) predicted geoid by M_1. model.

Because viscosity is thought to depend strongly on temperature. For the M_3 model providing the best fit, it show that the low viscosity in the uppermost mantle seem to be required beneath Tibetan Plateau, and we would deduce the Tibetan Plateau region is underlain by warmer mantle in the upper 400 km of the mantle.

Acknowledgments

This work was supported by the Laboratory of Geodynamic Geodesy, CAS and National Science Foundation of China under grant No. 49604057.

160

REFERENCES

1. G.B. Airy. On the computation of the effect of the attraction of the mountain-masses as disturbing the apparent astronomical latitude of stations in geodetic surveys, Phi. Trans. R. Soc. London. 145, 101-104 (1855).

2. C.G. Chase. The geoid effect of compensated topography and uncompensated oceanic trenches, Geophys. Res. Lett. 9, 29-30 (1982).

3. P.W. Forsyth. Surface loading and estimates of the flexural rigidity of continental lithosphere, J. Geophys. Res. 90, 12621-12632 (1985).

4. H.T. Hsu and Y. Lu. The regional Geopotential model in China , Bollettino Di Geodesia E Scienze Affini , n2, 161-175 (1995).

5. P. Karki, L.Kivioja and W.A. Heiskanen. Topographic-isostatic reduction maps for the world for the Hayford zone 18-1, Airy -Heiskanen system, T=30km. Helsinki, Publications of the isostatic institute of the international association of geodesy, no. 35, 239 (1961).

6. M.G. Kogan and M.K. McNutt. Gravity field over northern Eurasia and variation in the strength of the upper mantle, Sciences, 259, 473-478 (1993).

7. H. Lyon-Caen and P. Molnar. Constraints on the structure of the Himalaya from an analysis fo gravity anomalies and a flexural model of the lithosphere, J. Geophys. Res. 88, 8171-8192 (1983).

8. H. Lyon-Caen and P. Molnar. Gravity anomalies and the structure of the western Tibet and the southern Atrium basin, Geophys. Res. Lett. 11, 1251-1254 (1984).

9. M.K. McNutt, M. Diament and M. G. Kogan. Variation of elastic plate thickness at continental thrust belts, J. Geophys. Res. 93, 8825-8838 (1988).

10. R.L. Parker . The rapid calculation of potential anomalies , Geophys. J. R. 31, 447-455, 1972.

11. J.H. Pratt. On the attraction of the Himalaya Mountains and of the elevated regions beyond them, upon the plum line in India. Phil. Tran. R. Soc. London. 145, 53-100 (1855).

12. M.A. Richards and B.H. Hager , Geoid anomalies in a dynamic earth , J. Geophys. Res. 89, 5987-6002 (1984).

13. R.W. Simpson, R.C. Jachens and R.J. Blakely. AIRROOT: A FORTRAN program for calculating the gravitational attraction of an Airy isostatic root out to 166.7 km: U.S. Geological Survey Open-file Rept. 1-83 (1983).

14. R.W. Simpson, R.C. Jachens, R.J. Blakely and R.W. Saltus. A new isostatic residual gravity map of the conterminous United States with a discussion on the significance of isostatic residual anomalies, J. Geophys. Res. 8348-8327 (1986).

15 W. Su , R.L. Woodward and A.M. Dziewonski. Degree 12 model of shear velocity heterogeneity in the mantle, J. Geophys. Res. 6945-6980 (1994).

16. C.C.Tscherning and R. H. Rapp. Closed covariance expressions for gravity anomalies, geoid undulations and deflections of the vertical implied by anomaly degree variance model , Rept. 208, Dept. of Geod. Sci. and Surv. , Ohio State University (1974)

17. M. Wang and Z. Cheng , Isostatic gravity anomaly and curstal structure , ACTA Geologica Sinica , 51-61 (1981).

Proc. 30 *Int'l. Geol. Congr.*, Vol. 4, pp. 161-170
Hong Dawei (Ed)
© VSP 1997

Australian Continental-Scale Lineaments: Evidence of Reactivation of Deep-Seated Structures and Economic Implications.

CATHERINE I. ELLIOTT[1], INGRID CAMPBELL[2], E.B. JOYCE[3] AND C.J. WILSON[3]

[1]*Consultant, Minoil Exploration, PO Box 325, Wandin, 3139. Australia.*
[2]*Consultant, Continental Resources, 197 Canterbury Rd, Canterbury, 3126. Australia.*
[3]*School of Earth Sciences, University of Melbourne, Parkville, 3052. Australia.*

Abstract

The Australian continent is transected by a network of continental-scale lineaments that can be correlated with aligned geological, geophysical and geomorphological lithospheric expression. The continental-scale corridors are concentrated zones of aligned lineament distribution which manifest as corridors at a continental scale but can be broken down into individual, linear segments. This process of segmentation can be seen progressively, from the continental-scale down to the field scale.

The G3 and G5 gravity corridors in northwestern Australia are examples of ancient lineaments that have undergone a coupled reactivation history from the earliest Proterozoic through to the Holocene. The G5 coincides with the Proterozoic Halls Creek Mobile Zone and its lateral equivalent, the Fitzmaurice Mobile Zone. The G5 corridor can be traced offshore into the Arafura Basin where it has influenced the tectono-depositionary history of this sedimentary sequence. The G3 correlates with the Proterozoic King Leopold Mobile Zone in northwestern Australia and can be traced for a distance exceeding 3000km to the east coast of Australia.

Major Australian mineral and hydrocarbon occurrences are preferentially located at the intersections of continental-scale lineaments. Lineaments reflect basement architecture and form fluid-flow pathways within the lithosphere. Lineament intersections are zones of enhanced favourability in terms of geological structuring and, under suitable geochemical conditions, focus the transport, precipitation and entrapment of hydrocarbons and mineralized fluids.

Keywords: Lineaments, Basement Tectonics, Reactivation, Australia, Kimberley

INTRODUCTION AND HISTORICAL ASPECTS

The identification of large-scale lineaments and their relationship to the regional structuring of the Australian continent was first undertaken by pioneering geologist Edwin Sherbon Hills [7]. Hills recognized that the Australian basement architecture was intimately associated with a systematic network of continental-scale morphotectonic lineaments. He described these lineaments as ' zones of yielding' in the earth's crust and suggested that they acted as preferential pathways for ore-bearing fluids. Much of his morphotectonic work was based on the use of the ' Great Relief Model of Australia, a scaled topographic model which the Australian Government commissioned Hills to build for defence purposes in 1942. Hills observed that major mineral accumulations tended to be preferentially located along regional morphological lineaments and, more specifically, at the lineament intersections.

The work of Hills was further developed by the detailed lineament investigations of O'Driscoll [16], whose extensive publications on the subject date back to early 1960'. O'Driscoll has clearly demonstrated the lineament–ore relationship and his lineament-based

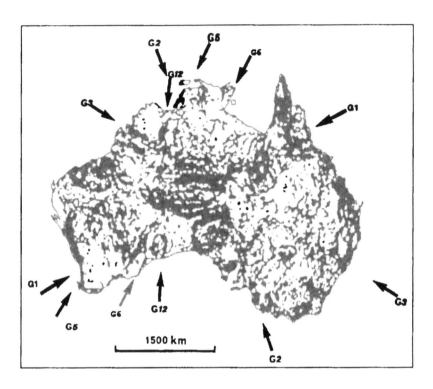

Figure 1. Diffused Bouger gravity map of Australia (O'Driscoll, 1980) showing several of the major gravity lineaments that influence northwestern Australia. The lineaments are best observed at a low angle along their strike.

tectonic models were fundamental to the discovery of the Olympic Dam Copper–Gold–Uranium (Cu-Au-U) deposit at Roxby Downs in 1975 [19,15]. Figure 1 is a diffused Bouger Gravity image of Australia, constructed by O'Driscoll through oscillating the negative during the development stage of the photographic process. This produced a blurred effect similar to modern computer imaging and still remains one of the most useful images for observing several of the continental-scale lineaments. He often referred to these as corridors because lineaments appeared as zones along which two parallel edges could be identified.

A spatial association between lineaments and hydrocarbons was shown by Campbell [1] along the north-northwest Port Campbell–Netherby Trough lineament in southern Australia, and the Cooper-Eromanga Basin of central eastern Australia [2,14]. North American explorationists such as Norman [11] and Foster [6] have demonstrated that lineament tectonics can be used to interpret basement structures associated with potential and proven hydrocarbon reserves. This paper provides an outline of some of the results of

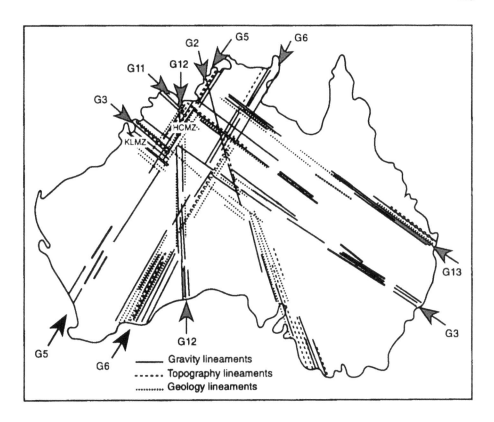

Figure 2. A schematic diagram showing the concentration of aligned lineament segments along several of the continental-scale gravity corridors. The corridors are zones of concentrated gravity, magnetic and topographic alignments giving the appearance of continuity over vast distances, often in the order of thousands of kilometres.

the author's PhD investigation into the lineament tectonics of northwestern Australia studied in the broader context of the Australian framework.

THE AUSTRALIAN LINEAMENT FRAMEWORK

The Australian continent is transected by a network of systematic continental-scale lineaments that have undergone a series of reactivation events that date back to the earliest Proterozoic. O'Driscoll published a sequence of 10 major gravity corridors [13] each being hundreds to thousands of kilometres in length and interpreted as continuous zones. Hills [7] on the other hand, chose to portray the regional lineaments as segments which he tentatively joined in recognition of their aligned nature over vast distances. The author's studies have indicated that the continental-scale corridors are concentrated zones of aligned lineament distribution (Fig. 2) and that though manifest as corridors at a continental scale, can be broken down into individual, linear segments. This process of segmentation can be seen progressively, from the continental-scale down to the field scale, a consequence of the scale invariant nature of lineaments [8], producing a ' lineaments within lineaments' effect.

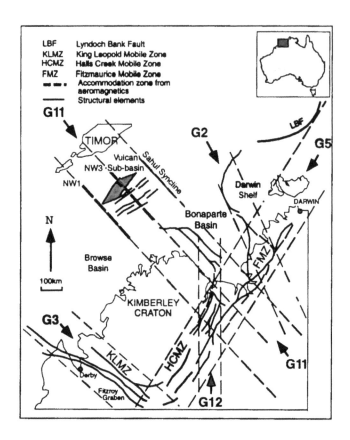

Figure 3. Regional tectonic elements of northwestern Australia showing the coincidence of major structural trends with gravity interpreted continental scale lineaments.

A directional frequency rose diagram plot of 545 large-scale lineaments interpreted from the numerous data sets (3,4), shows a preferential distribution concentrated along the orthogonal pair of west-northwest to northwest and north-northeast to northeast with a secondary orientation of north-south. The north-northeast Halls Creek Mobile Zone (HCMZ) and west-northwest King Leopold Mobile Zone (KLMZ) in northwestern Australia are an example of such a pair. They have behaved as a coupled, orthogonal system that has undergone a complex reactivation history from the earliest Proterozoic through to the Holocene [18].

EXAMPLES OF REACTIVATION AND LONGEVITY OF CONTINENTAL-SCALE LINEAMENTS

The G3 and G5 corridors are clearly seen using O'Driscoll's [15] diffused gravity image (Fig. 1). In northwestern Australia, these lineaments align with major tectonic elements and depocentre margins of both Pre-Cambrian and Phanerozoic age (Fig. 3). Several examples are given below that demonstrate that these large-scale lineaments were active in the Proterozoic and, through reactivation, have influenced the Phanerozoic tectono-depositional history, in some cases to the present day.

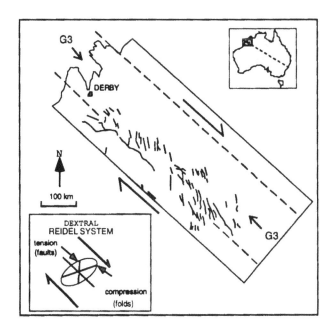

Figure 4. Ingredient diagram showing Permo–Triassic faulting of the northern Canning Basin is concentrated along the southern edge of the G3 gravity corridor. The NS orientation indicates dextral shear along the WNW G3 during this period.

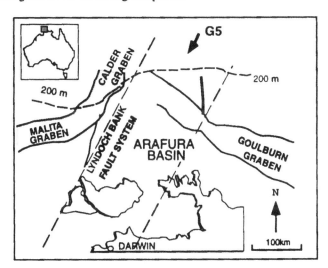

Figure 5. Schematic tectonic elements diagram of the offshore Arafura Basin. The NE Malita Graben terminates against the western edge of the G5 where it swings into a NNE trend becoming the Calder Graben (Modified from McLennen et al., 1990).

The G3 Corridor

The west-northwest G3 corridor is a continental-scale zone comprising concentrated west-northwest lineaments that O'Driscoll [15] and Elliott [3,4,5] have interpreted to extend from northwestern Australia to the eastern coastline near Sydney (Fig. 1). In northwestern Australia, the corridor coincides with the KLMZ (Fig. 3), a reactivated mobile belt comprising rifting, wrenching, thrusting and concentrated igneous activity over a period in excess of 1800 Ma [18]. During its reactivation history, the G3 has behaved as a transfer zone to the Late Palaeozoic–Mesozoic rifting of Gondwana along the northwest margin of the continent [18]. The distribution of Permo-Triassic faulting supports a dextral sense of shear under a reidel system during this period (Fig. 4). The faulting is notably concentrated within the gravity interpreted corridor of the G3.

The G5 Corridor

The G5 corridor is a major gravity discontinuity that can be traced as a continuous north-northeast zone through northwestern and northern Australia (Fig.1). The corridor coincides with the HCMZ (Fig. 3) and its lateral equivalent, the Fitzmaurice Mobile Zone (FMZ), both ancient zones of repeated basement reactivation [17,18]. The linear, eastern margin of the Bonaparte Basin approximates the eastern edge of the G5 corridor. Further north and offshore in the Arafura Basin, the G5 corridor can be identified as a double-edged basement discontinuity in the aeromagnetics [3,4]. The western edge aligns with the Lyndoch Bank Fault Zone along the edge of the offshore Arafura Basin (Fig. 5), an ancient fault zone with a reactivation history dating back to the Pre-Cambrian [10]. An offset in the bathymetric 200 m contour suggests a very recent influence of the basement discontinuity. Seasat imagery and regional tectonics have been used to trace the influence of the north-northeast basement discontinuity across the Australian Plate, possibly as far as Irian Jaya [5].

ECONOMIC IMPLICATIONS

Given favourable geochemical conditions, anomalous mineral and hydrocarbon occurrences tend to be concentrated along lineament edges with the major discoveries located at lineament intersections. A combined plot of several major Australian mineral deposits and continental-scale lineaments (Fig. 6), demonstrates the recurring spatial association between the deposits and the lineament intersections. Major Australian petroleum provinces show a similar distribution, being preferentially located at the intersections (Fig. 7).

In northwestern Australia major mineral and hydrocarbon occurrences also coincide with the regional lineament intersections (Fig. 8). The NS1 lineament is interpreted onshore using satellite imagery; and offshore, it aligns with a major north-south aeromagnetic lineament in the Vulcan Graben [12]. The intersection zone of the NS1 with the northwest G11 [3,4] is coincident with the Jabiru, Challis, Skua, Puffin and Montara oilfields. Onshore, the NS1 / G3 intersection is coincident with the only significant Canning Basin oil discoveries, together with the Ellendale and Noonkanbah diamondiferous lamproites. The Mississippi style Pb-Zn deposits of Cadjebut and Blendevale are located at the intersection of the G3 with the G5 corridor, the world class Argyle diamond deposit is at the G5, G11 and G12 intersection.

At a more local scale, major mineral deposits of the Halls Creek region in the southeastern Kimberley (G3 / G5 intersection) are spatially associated with regional structural corridors identified from lineament analyses of satellite imagery [3,4]. It was shown that the crosscutting regional structural corridors (identified from the satellite data), and annotated

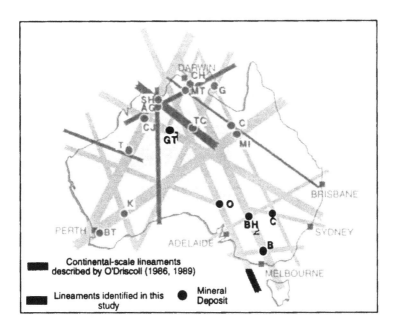

Figure 6. Schematic diagram showing the spatial association between major mineral accumulations and continental lineaments, particularly lineament intersections. **AG** = Argyle diamonds; **B** = Bendigo Au; **BH** = Broken Hill Ag-Pb-Zn; **C** = Cobar Cu-Au; **CJ** = Cadjebut Pb-Zn; **CH** = Coronation Hill Au; **G** = Groote Eylandt Mn; **K** = Kalgoorlie Au/Kambalda Ni; **MI** = Mt Isa Pb-Zn; **O** = Olympic Dam Cu-Au-U; **T** = Telfer Au.

HC1, HC2 and HC3, as zones of preferential lineament alignment. In the field, the northeast HC1 corridor is orientated at a low angle to the north-northeast trend of the HCMZ (G5) and is denoted by a zone of parallel faults, drainage channels, a northeast trending mountain range, and a concentration of the Early Proterozoic granites. The cross-cutting HC2 and HC3 are less well-defined and could easily be missed without the aid of satellite imagery. They are predominantly expressed in the field by aligned faults. The two largest deposits Koongie Park Pb-Zn-Cu, and Sally Malay Ni-Cu [9], are found at the intersections of the regional lineaments.

CONCLUSIONS

Concentrated zones of aligned lineaments are recognized as continental-scale corridors that in many cases appear to form continuous zones across the country. The antiquity of many is verified by their expression in the earliest Proterozoic; their longevity, by their documented reactivation through time. Studies have shown that these continental-scale lineaments are aligned zones of concentrated tectonic, igneous and depositional activity, and have controlled the regional tectonic and depositional histories along individual corridors such as the G3 and G5. The preferential alignment of basin margins suggests that fault activity concentrated along, and parallel to, the lineaments has acted as hinge zones to the regional depocentres.

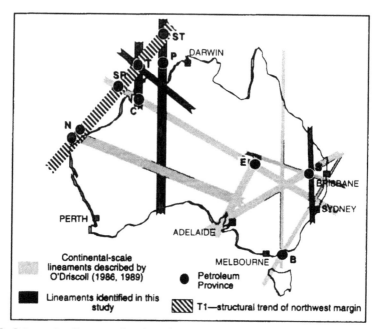

Figure 7. Schematic diagram showing the spatial association between major petroleum provinces and continental lineaments, particularly lineament intersections. B = Bass Strait; C = Canning Basin; E = Cooper-Eromanga; N = Northwest Shelf; P = Petrel/Tern; SR = Scott Reef; ST = Sunrise/Troubadour; T = Timor Sea.

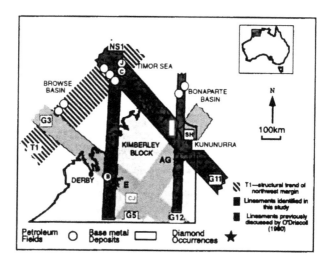

Figure 8. Schematic diagram showing the coincidence of major mineral and hydrocarbon occurrences in northwestern Australia with the intersection zones of the continental-scale lineaments. **AG** = Argyle diamonds; **B** = Blina oil; **C** = Challis oil; **CJ** = Cadjebut Pb-Zn; **E** = Ellendale diamonds; **J** = Jabiru oil; **N** = Noonkanbah diamonds; **SH** = Sorby Hills Pb-Zn.

The spatial association between lineaments and mineral and hydrocarbon occurrences can be recognized at all scales, from the continental down to the local. The fundamental input of lineament tectonics into the discovery of the Olympic Dam Cu-Au-U deposit is testimony to the enormous potential of the technique to the exploration process. At all scales, given favourable geology and geochemistry, resources appear to be preferentially associated with lineaments and, in particular, lineament intersections.

Given that lineaments are zones of energy release (i.e. igneous activity, faulting etc., and zones of fluid flow), intersections appear to be favoured sites for the formation or entrapment of resource commodities. This may be because intersections have been tectonically active zones through time. As such they are zones of increased fracture porosity, conduits for deep crustal and mantle fluids, and potential mixing zones for fluids of differing chemistries. In hydrocarbon exploration, lineament intersections will be areas of increased fracture porosity (migration pathways), tectonically active sites of erosion and deposition (reservoir formation), areas of higher heat flow (maturation) and zones of enhanced structural activity (trap formation).

Acknowledgements

The author gratefully acknowledges the following companies for their financial support of the Ph.D. study: BHP Petroleum, BHP Minerals, ACS Laboratories, Western Mining Corporation, Acacia Resources and Stockdale Prospecting. Thanks are due also to my Ph.D. supervisors, I. B. Campbell, E.B. Joyce and C.J. Wilson.

REFERENCES

1. Campbell, I.B., 1989. The Port Campbell–Netherby Trough structural corridor in southeastern Australia. In: LeMaitre (Ed) *Pathways in Geology: Essays in honour of Edwin Sherbon Hills*. Hills Memorial Volume Committee, pp 280-303
2. Campbell, I.B., and O'Driscoll, E.S.T., 1989. Lineament–hydrocarbon associations in the Cooper and Eromanga Basins. In: O'Neil, B.J., (Ed.) *The Cooper and Eromanga Basins Symposium*. pp 295-313
3. Elliott, C.I., 1994a. Australian lineament tectonics with an emphasis on northwestern Australia. *Ph.D. Thesis. University of Melbourne* (Unpubl.) 264 pp.
4. Elliott, C.I., 1994b. Lineament tectonics: An approach to basin analysis and exploration. In Purcell, P. and R., *Sedimentary Basins of Western Australia (Conference Proceedings). Petroleum Exploration Society Australia*, pp 77–90.
5. Elliott, C.I., 1994c A new investigation of some Australian gravity lineaments. *Bulletin Geological Society Malaysia.* 33, pp 357-368.
6. Foster, N.H., 1978. Geomorphic exploration used in the discovery of Trap Spring oilfield, Railroad Valley, Nye County, Nevada. *American Association Petroleum Geologists Bulletin* (Abstract). 62(5). p 884
7. Hills, E.S., 1956. A contribution to the morphotectonics of Australia. Benchmark paper reprinted In: LeMaitre (Ed) *Pathways in Geology: Essays in honour of Edwin Sherbon Hills*. Hills Memorial Volume Committee, pp 230-246.
8. Hobbs, B., 1991. Interfacing structural geology, ore reserves and rock mechanics in the mine environment. Extended abstract. *Proceedings of the structural geology in mining and exploration conference*. University of Western Australia Geology Department. 25. pp 140-141
9. Marston, R.J., Blight, D.F and Morrison, R.J., 1981. Mineral deposits of Western Australia. 1:2,500,000 map. *Geological Survey of Western Australia.*
10. McLennen, J.M., Rasidi, J.S., Holmes, R.L. and Smith, G.C., 1990. The geology and petroleum potential of the western Arafura Sea. *Journal Geological Society of Australia*, 32(3), pp 272-285.

11. Norman, J.W., 1976. Photogeological fracture trace analysis as a subsurface exploration technique. *Transactions Mining and Metallurgy.* Section B. 85. pp 52-61
12. O'Brien, G.W., 1993. Some ideas on the rifting history of the Timor Sea from the integration of deep crustal seismic and other data. *PESA Journal.* 21, pp 95-115
13. O'Driscoll, E.S.T., 1980. The double helix in global tectonics. *Tectonophysics.* 63. pp 397-417
14. O'Driscoll, E.S.T., 1983. Deep tectonic foundations of the Eromanga Basin. *APEA Journal.* 23(1), pp 5-17
15. O'Driscoll, E.S.T., 1985. The application of lineament tectonics in the discovery of the Olympic dam Cu-Au-U deposit at Roxby Downs, South Australia. *Global tectonics and metallogeny.* 3 (1). pp 43-57
16. O'Driscoll, E.S.T., 1990. Lineament tectonics of Australian ore deposits. In: Hughes (Ed) *Geology of the mineral deposits of Australia and Papua New Guinea.* Aus.I.M.M., pp 33-41
17. Plumb, K. A., 1990. Halls Creek province and the Granites-Tanami inlier—regional geology and mineralization. In: Hughes (Ed) *Geology of the mineral deposits of Australia and Papua New Guinea.* Aus.I.M.M., pp 681-687.
18. White, S.H., and Muir, M.D., 1989. Multiple reactivation of coupled orthogonal fault systems: An example from the Kimberley region in northwestern Australia. *Geology.* 17. pp 618-621
19. Woodall, R., 1994. Empiricism and concept in successful mineral exploration. *Australian Journal of Earth Sciences.* 41(1). pp 1-10

9 780367 448103